INTRODUCTION TO NUMBER THEORY

INTRODUCTION TO NUMBER THEORY

Daniel E. Flath
Department of Mathematics
National University of Singapore

A WILEY-INTERSCIENCE PUBLICATION
JOHN WILEY & SONS
NEW YORK • CHICHESTER • BRISBANE • TORONTO • SINGAPORE

Library of Congress Cataloging-in-Publication Data

Flath, Daniel E.
Introduction to number theory.

"A Wiley-Interscience publication."
Bibliography: p.
1. Numbers, Theory of. I. Title.
QA241.F59 1989 512'.7 88-10638
ISBN 0-471-60836-X

Printed in the United States of America

10 9 8 7 6 5 4 3 2 1

To Laura

Preface

When I came to Singapore I was given a fourth year undergraduate Honours Number Theory course. I decided to teach Gauss's immortal *Disquisitiones Arithmeticae*. This book is the result.

On historical and mathematical grounds alike number theory has earned a place in the curriculum of every mathematics student. This is a textbook for an advanced undergraduate or beginning graduate core course in the subject. Such a course should stick pretty close to the naive questions, which in number theory concern prime numbers and Diophantine equations. The emphasis in this book is on Diophantine equations, especially quadratic equations in two variables.

My own conscious interest in Diophantine equations goes back to a long winter's night in a St. Louis basement in 1962 when my father and I tried to solve the notorious problem of the monkey and the coconuts as presented by Martin Gardner. No one told me then that Diophantine equations belong to a subject called "number theory," and I found little help in the public library. I needed a teacher trained in number theory. It pleases me that several of my students of Gauss are now teaching in the schools. I might particularly mention Mr. Lee Ah Huat with whom I discovered Gauss's first proof of the law of quadratic reciprocity.

This book is closely based on lectures I gave to able groups of students during three consecutive years at the National University of Singapore. I thank the students for constantly demanding "the notes," which was how the text began. I tried during the writing always to keep my students in mind, always to remember that I was writing a textbook. I have sought to avoid the twin traps of doing algebra to the exclusion of number theory and of doing only trivial number theory. I take it for granted that the material I have chosen is interesting. My supreme stylistic goal is clarity.

By the time this book is published I shall have gone on from Singapore. Singapore has been part of my life for three and a half years, and I shall miss it. I have many friends here. I wish them all well.

Singapore
August 1987

DAN FLATH

Acknowledgment

I am very grateful to Jean-Pierre Serre and the Singapore Mathematical Society for permission to reprint the article "$\Delta = b^2 - 4ac$," which was first published in the *Mathematical Medley*.

DANIEL E. FLATH

Contents

CHAPTER 1

Prime Numbers and Unique Factorization

1. Introduction

All introductions to number theory since Gauss's greatest *Disquisitiones Arithmeticae* (1801) have begun the same way, with the aptly named Fundamental Theorem of Arithmetic. What better place to start?

Theorem 1.1. Fundamental Theorem of Arithmetic. Every integer greater than 1 can be expressed as a product of prime numbers in one and only one way.

Thus he who would know the integers well must study the primes. The first question to ask is just how many primes are there? We shall soon see that there are infinitely many primes, so we had better refine the question. We will prove

Theorem 1.2. Chebyshev. For $x > 2$, the number of primes that are less than x is between $(1/10)(x/\log x)$ and $10(x/\log x)$. That is, to within an order of magnitude there are $x/\log x$ primes less than x.

We prove Theorem 1.1 in Section 3. The proof will depend upon properties of greatest common divisors that we will use again and again. A first application to linear Diophantine equations is the subject of Section 4.

There is a preliminary discussion of the distribution of primes in Section 2. A proof of Theorem 1.2 is presented in Section 5.

2. Prime Numbers

Divisibility is without a doubt the most important concept in number theory.

Definition. A nonzero integer b is said to *divide* $a \in \mathbb{Z}$ (written $b|a$) iff there exists $c \in \mathbb{Z}$ such that $a = bc$. We say then that b is a *divisor* of a.

Definition. A positive integer n is *prime* iff $n > 1$ and there is no factorization $n = ab$ with positive integers a, $b < n$.

A positive integer n is *composite* iff $n > 1$ and n is not prime.

Note that a positive integer greater than 1 is prime if and only if its only positive divisors are itself and 1.

Every positive integer can be constructed from prime numbers by multiplication. This fact, the first theorem in the subject of number theory, is the source of our interest in primes.

Theorem 2.1. Every positive integer $n > 1$ is a product of prime numbers: $n = \prod_{i=1}^{r} p_i$, $r \geq 1$.

Proof. By finite induction on n. If $n = 2$, the assertion is true because 2 is prime.

Let $n > 2$ and assume that every positive integer less than n is a product of primes. If n itself is prime, the assertion of the theorem is clearly true. If n is composite, write $n = ab$ where $1 < a, b < n$. By the inductive hypothesis a and b are each products of primes. Juxtaposing these products gives a prime factorization of n. ■

How many primes are there? The answer was known to Euclid.

Theorem 2.2. Euclid. There are infinitely many primes.

Proof. We will show that every finite set of primes omits at least one prime. It will follow that no finite set can contain all the primes.

Let $\{p_1, p_2, \ldots, p_r\}$ be a finite set of prime numbers. By Theorem 2.1, there is a prime divisor q of $N = p_1 p_2 \cdots p_r + 1$. Because $q|N$ but $p_i \nmid N$, the prime q must be different from $p_1, p_2, \ldots, p_r$. ■

The occurrence of primes is limited somewhat by the existence of long strings of consecutive composite numbers.

Proposition 2.3. There exist N consecutive composite integers for every $N \geq 1$.

Proof. We can take, for example, the sequence $(N+1)!+2, (N+1)!+3, (N+1)!+4, \ldots, (N+1)!+(N+1)$. None of these numbers is prime because i divides $(N+1)!+i$ for $2 \leq i \leq N+1$. ■

Some theorems of a nature opposite to Proposition 2.3 have been discovered. For instance, Chebyshev proved that every sequence $N+1, N+2, \ldots, 2N$ contains at least one prime. It has been conjectured that the sequences $N^2+1, N^2+2, \ldots, (N+1)^2$ each contain at least one prime too, but this has never been proved and no counterexample is known.

By definition, a positive number n is prime if it is not divisible by any of the integers $2, 3, \ldots, n-1$. It has been known since Eratosthenes, however, that the number of divisibility checks can be drastically reduced, which greatly shortens the work of compiling lists of primes.

Proposition 2.4. Sieve of Eratosthenes. Let n be a composite number. Then there exists a prime p such that $p|n$ and $p \leq \sqrt{n}$.

Proof. Since n is composite, $n = ab$ with $1 < a, b < n$. If $a, b > \sqrt{n}$, then $ab > n$. Hence at least one factor, say a, satisfies $a \leq \sqrt{n}$. Every prime divisor p of a will be as desired: $p|n$ and $p \leq \sqrt{n}$. ■

Let us find the primes smaller than 60. Begin with a list of the first 60 integers. The primes smaller than $\sqrt{60}$ are 2, 3, 5, and 7. So we strike from the list, in turn: 1, every second number beginning with 4, every third number beginning with 9, every fifth number beginning with 25, and finally every seventh number beginning with 49. Those numbers that remain, 17 in all, are the primes less than 60. They are circled in Table 1.

The arrangement of Table 1, imagined now to continue indefinitely to the right, suggests some questions about the distribution of primes. By Theorem 2.2 we know that an infinite number of columns of the extended table will

TABLE 1

1	(5)	9	(13)	(17)	21	25	(29)	33	(37)	(41)	45	49	(53)	57
(2)	6	10	14	18	22	26	30	34	38	42	46	50	54	58
(3)	(7)	(11)	15	(19)	(23)	27	(31)	35	39	(43)	(47)	51	55	(59)
4	8	12	16	20	24	28	32	36	40	44	48	52	56	60

contain a prime. Is there an infinite number of columns containing two primes? No one knows the answer.

The infinite set of odd primes divides itself between the first and third rows of Table 1. At least one of those rows must contain an infinite number of primes. Which is it? Observe that the ith row of Table 1 consists of the integers $i + 4k$, $k = 0, 1, 2, \ldots$. We have:

Theorem 2.5. There is an infinite number of integers $k \geq 0$ such that $3 + 4k$ is prime.

Proof. We can imitate the proof of Theorem 2.2. Let $p_1, \ldots, p_r$ be a list of prime numbers of the form $3 + 4k$. Let $N = 4p_1 \cdots p_r - 1$. Since N is odd, all its prime divisors lie in rows 1 and 3 of Table 1; that is, each of its prime divisors is of the form $1 + 4k$ or $3 + 4k$. An easy exercise shows that every product of integers of the form $1 + 4k$ is itself also of the form $1 + 4k$. Since N is not of this form, it must have a prime divisor q of the form $3 + 4k$. Because $q | N$ but $p_i \nmid N$, the prime q must be different from $p_1, \ldots, p_r$. We have shown that every finite list of primes of the form $3 + 4k$ is incomplete. ■

Thus the third row of Table 1 contains infinitely many prime numbers. The first row also contains an infinite number of primes, a fact whose proof will be deferred until after the discussion of some elementary properties of congruences.

It is natural now to ask, for given a and b, whether there is an infinite number of integers X for which $aX + b$ is a prime number. An obvious necessary condition is that a and b have no prime divisor in common, for if a prime p divides both a and b, then it will also divide $aX + b$ for every X. One of the mathematical high points of the nineteenth century was Dirichlet's proof, introducing powerful new analytic methods into number theory, that this necessary condition is also sufficient. We state without proof Dirichlet's Theorem on Primes in Arithmetic Progressions.

Theorem. Dirichlet. Let $a, b \in \mathbb{Z}$ with $a \geq 1$. If there is no prime that divides both a and b, then $aX + b$ is a prime number for infinitely many positive integers X.

What if we replace the linear polynomial $aX + b$ by a polynomial in X of higher degree? Absolutely nothing is known, not even for the simplest case, which is the polynomial $X^2 + 1$. It is suspected but not proved that $X^2 + 1$ is prime for an infinite number of integers X.

Polynomials in more than one variable are interesting too. Fermat, in the seventeenth century, created modern number theory by considering the polynomial $X^2 + Y^2$. He proved that the equation $X^2 + Y^2 = p$ is solvable in integers X, Y for every prime number p that is of the form $4k + 1$ (row 1 of Table 1) and for no prime number p of the form $4k + 3$ (row 3 of Table 1). Thus there is an infinite number of primes of the form $X^2 + Y^2$. We will present several proofs of Fermat's theorem later. Following the historical path, we will then take up the theory of the general binary quadratic form $aX^2 + bXY + cY^2$.

Exercises

1. **i.** Show that the set of integers $S = \{1, 5, 9, 13, \dots\}$ is a multiplicative set (i.e., that the product of any two elements of S lies in S).

ii. An *S-prime* is a number in S other than 1 that is not the product of two smaller numbers in S. Prove that every element of S is a product of S-primes but that such a factorization is not always unique.

2. Prove that if $2^n + 1$ is prime, then n is a power of 2. Compute $2^{2^n} + 1$ for $n = 0, 1, 2, 3, 4$. Are they prime?

3. Show that if $m < n$, then $2^{2^m} + 1$ divides $2^{2^n} - 1$ and so there is no prime number that divides both $2^{2^m} + 1$ and $2^{2^n} + 1$. Conclude that there are infinitely many prime numbers.

4. Prove that if $M_p = 2^p - 1$ is prime, then p itself is prime. Compute M_p for $p = 2, 3, 5, 7, 11$. Are they prime? (The largest prime known in 1988 is M_{216091}.)

5. Prove that there are infinitely many primes of the form $3k + 2$. Prove that every such prime except 2 is also of the form $6k + 5$.

6. Deduce from Dirichlet's theorem that there are infinitely many primes with final digit any of $1, 3, 7, 9$. Show similarly that there are infinitely many primes whose final two digits are 37.

7. Find all primes less than 200 by hand. Using a computer, list all primes less than 1000. How many of these primes have final digit 1? final digit 3? 5? 7? 9? Find all primes $p < 1000$ such that $p + 2$ is also prime. For which of these p is $p + 4$ also prime? For which is $p + 6$ also prime? For which are $p + 6$ and $p + 8$ both prime?

8. On the basis of experiment, formulate a conjecture telling for which prime numbers p there exist integers X and Y such that $X^2 + 3Y^2 = p$. Do the same for the equation $X^2 + XY - Y^2 = p$.

9. Prove that there is no nonconstant polynomial $f(X)$ with integer coefficients such that $f(n)$ is prime for every positive integer n.

10. Let N be a positive odd integer. Let $x = \inf\{n \in \mathbb{Z} | n \geq \sqrt{N}$ and $n^2 - N$ is the square of an integer$\}$. Let $y = \sqrt{x^2 - N}$. Prove that $x - y$ is the largest divisor of N that is less than or equal to $\sqrt{N}$.

3. Unique Factorization

It was discovered in ancient times that the set of common divisors of any two integers is equal to the set of all divisors of a single third integer, called their greatest common divisor. The greatest common divisor can be calculated by a very efficient procedure, today known as the Euclidean algorithm. This algorithm is the key to understanding factorization in the ring of integers. Our proof that factorization of integers into products of primes can be done in only one way, the principal result of this chapter, will be based solidly upon it.

This story begins with the possibility of division with remainder.

Proposition 3.1. Division Algorithm. Let $a, b \in \mathbb{Z}$ with $b \neq 0$. There exist $q, r \in \mathbb{Z}$ such that $a = qb + r$ and $|r| < |b|$. (That is, we can divide a by b, getting quotient q with a remainder r which is smaller than the divisor b.)

Proof. Let q be the largest integer that is less than or equal to the rational number a/b. From the inequalities $q \leq a/b < q + 1$, we deduce that $|a/b - q| < 1$. Let $r = a - qb$. Note that r is an integer because a, q, and b are integers. Finally, compute $|r| = |b(a/b - q)| = |b|\,|a/b - q| < |b|$. ■

Definition. An integer c is a *common divisor* of two integers a and b iff c divides both a and b. It is a *greatest common divisor* (GCD) of a and b iff it is a common divisor of a and b and is itself divisible by all other such common divisors.

Proposition 3.2. Euclidean Algorithm. There is a greatest common divisor for every pair of integers not both zero. It can be computed by a finite number of applications of the division algorithm.

Proof. Let a, b be integers with $b \neq 0$. If $b|a$, then b is a GCD of a and b. If not, write inductively, using Proposition 3.1,

$$\begin{aligned} a &= q_1 b + r_1, \\ b &= q_2 r_1 + r_2, \\ r_1 &= q_3 r_2 + r_3, \\ &\vdots \\ r_{m-1} &= q_{m+1} r_m + r_{m+1}, \qquad m = 3, 4, \ldots, \end{aligned}$$

where $q_i, r_i \in \mathbb{Z}$ and $|b| > |r_1| > |r_2| > \cdots$. Because the $|r_i|$ are nonnegative integers, the process must stop with an integer $M \geq 1$ such that $r_{M+1} = 0$. We claim that r_M is a GCD for a and b.

From the equation $a = q_1 b + r_1$, it is evident that every common divisor of a and b must also divide r_1 and that every common divisor of b and r_1 must also divide a. Hence $S(a, b) = S(b, r_1)$, where we write $S(m, n)$ for the set of common divisors of two integers m, n. Consideration in turn of the succeeding equations above shows similarly that $S(b, r_1) = S(r_1, r_2)$ and that $S(r_m, r_{m+1}) = S(r_{m+1}, r_{m+2})$ for $m \geq 1$. Thus $S(a, b) = S(r_M, r_{M+1})$. But since $r_{M+1} = 0$ and every nonzero integer divides 0, $S(r_M, r_{M+1})$ is just the set of all divisors of r_M, including r_M. We conclude that the set of common divisors of a and b consists of r_M and the divisors of r_M. It is now clear that r_M is a GCD of a and b, as claimed. ■

In fact, every nonzero pair of integers has exactly two greatest common divisors, one positive and one negative. Indeed, if c_1 and c_2 are both GCDs of a and b, then $c_1|c_2$ and $c_2|c_1$, which shows that c_1/c_2 and c_2/c_1 are both integers. Hence $c_1/c_2 = \pm 1$ and $c_1 = \pm c_2$. We will denote the *positive* greatest common divisor of a and b by GCD(a, b). The integers a and b are said to be *relatively prime* iff GCD(a, b) = 1.

The usefulness of GCDs derives mainly from their most important property, which is stated as the next proposition.

Proposition 3.3. Let a, b be integers not both zero and let c be a greatest common divisor of a and b. Then there exist integers x, y such that $ax + by = c$.

Proof. It is enough to prove the proposition when c is any one of the two GCDs of a and b, for if $ax + by = c$, then $a(-x) + b(-y) = -c$.

Assume that $b \neq 0$. If $b|a$, then b is a GCD of a and b, and $b = a(0) + b(1)$.

Now suppose that $b \nmid a$, and adopt the notation of the proof of Proposition 3.2. We must prove that there exist integers $x, y \in \mathbb{Z}$ such that $r_M = ax + by$.

We show more generally that there exist $x_i, y_i \in \mathbb{Z}$ such that $r_i = ax_i + by_i$ for $i = 1, 2, \ldots, M$. The proof is by induction on i. The cases $i = 1$ and $i = 2$ come from the calculations

$$r_1 = a(1) + b(-q_1),$$
$$r_2 = b - q_2(a - q_1 b) = a(-q_2) + b(1 + q_1 q_2).$$

Now assume that $m > 2$ and that the existence of x_i, y_i has been proved for all $i < m$. Then

$$\begin{aligned} r_m = r_{m-2} - q_m r_{m-1} &= (ax_{m-2} + by_{m-2}) - q_m(ax_{m-1} + by_{m-1}) \\ &= a(x_{m-2} - q_m x_{m-1}) + b(y_{m-2} - q_m y_{m-1}). \end{aligned}$$

So taking $x_m = x_{m-2} - q_m x_{m-1}$ and $y_m = y_{m-2} - q_m y_{m-1}$ completes the induction. ■

Lemma 3.4. Euclid's Lemma. If a prime number divides a product $a_1 a_2 \cdots a_n$ of integers, then it must divide one of the a_j.

Proof. The proof is by induction on n, the number of factors in the product. If $n = 1$ there is nothing to prove.

The crucial case is that of $n = 2$, so let p be a prime divisor of a product ab. We must show that if $p \nmid a$, then $p | b$. Suppose that p does not divide a. Since 1 and p are the only positive divisors of p, we must have that $\mathrm{GCD}(p, a) = 1$. By Proposition 3.3 there exist $x, y \in \mathbb{Z}$ such that $ax + py = 1$. Multiply by b to get $(ab)x + pby = b$. Both terms on the left side of this equation are visibly divisible by p. Hence b, which is their sum, is also divisible by p.

Now let $N > 2$ and assume that the proposition has been proved for all $n < N$. Let p be a prime divisor of $(a_1 a_2 \cdots a_{N-1})a_N$. By the preceding paragraph, p must divide either a_N or $a_1 a_2 \cdots a_{N-1}$. In the latter case, the inductive hypothesis implies that p divides a_j for some $j \leq N - 1$. Hence, in all cases, p must divide one of the factors $a_1, a_2, \ldots, a_N$. ■

The Fundamental Theorem of Arithmetic, Theorem 1.1, can now be proved. It is the conjunction of our next result with Theorem 2.1.

Theorem 3.5. Uniqueness of Factorization. Apart from rearrangements of the prime factors, a positive integer can be expressed as a product of primes in only one way.

Proof. Let n be a positive integer. Suppose that $n = p_1 p_2 \cdots p_r = q_1 q_2 \cdots q_s$ where the p_i and q_j are all prime. We must show that $r = s$ and that after a possible reordering of the qs, $p_i = q_i$ for $i = 1, 2, \ldots, r$.

The proof is by induction on n. There is nothing to prove if $n = 1$. Let $n > 1$ and assume that the theorem has been proved for all positive integers less than n. If n is prime, then it has only one prime divisor, namely itself. hence $r = s = 1$ and $p_1 = q_1 = n$. Suppose that n is composite. The prime p_1 divides the product $q_1 q_2 \cdots q_s$, so by Euclid's lemma p_1 must divide q_j for some j. After rearranging the qs if necessary, we may assume that $p_1 | q_1$. Because q_1 is prime, $p_1 = q_1$. Thus $n/p_1 = p_2 \cdots p_r = q_2 \cdots q_s$. By the inductive hypothesis, n/p_1 has only one factorization as a product of primes. Thus $r = s$ and $p_i = q_i$ for $i = 2, \ldots, r$. ■

We conclude this section with two lemmas that we will need in the sequel. Their simple proofs illustrate nicely the use of Proposition 3.3.

Lemma 3.6. Let $a, b, m \in \mathbb{Z}$ with $m \neq 0$. If $m|ab$ and $\text{GCD}(m, a) = 1$, then $m|b$.

Proof. By Proposition 3.3 there exist $x, y \in \mathbb{Z}$ such that $ax + my = 1$. After multiplying by b, one finds that $(ab)x + mby = b$. Since m divides both terms on the left side of the equation, it must also divide b, which is their sum. ■

Lemma 3.7. Let $a, b \in \mathbb{Z}$, with a and b not both zero, and let $d = \text{GCD}(a, b)$. Then $\text{GCD}(a/d, b/d) = 1$.

Proof. By Proposition 3.3 there exist $x, y \in \mathbb{Z}$ such that $ax + by = d$. Thus $(a/d)x + (b/d)y = 1$. From this equation we see that all common divisors of a/d and b/d must divide 1. Hence, the only common divisors of a/b and b/d are ± 1. ■

Exercises

1. Show that the integers q and r of Proposition 3.1 are not necessarily uniquely determined by the pair (a, b).

2. Show that $\text{GCD}(2^e - 1, 2^f - 1) = 2^{\text{GCD}(e, f)} - 1$ for every pair of positive integers e, f.

3. Define the Fibonacci sequence by $F_0 = 0$, $F_1 = 1$, $F_n = F_{n-1} + F_{n-2}$ for $n \geq 2$.

   i. Prove that

$$\begin{pmatrix} F_{n+1} & F_n \\ F_n & F_{n-1} \end{pmatrix} = \begin{pmatrix} 1 & 1 \\ 1 & 0 \end{pmatrix}^n \quad \text{for } n \geq 1.$$

ii. Show that if $m > n$, then $\mathrm{GCD}(F_m, F_n) = \mathrm{GCD}(F_{m-n}, F_n)$.
iii. Prove that $\mathrm{GCD}(F_m, F_n) = F_{\mathrm{GCD}(m,n)}$.

4. Let $a, b, q_i, r_i \in \mathbb{C}$ with $b \neq 0$, all $q_i \neq 0$, $r_i \neq 0$ for $i \leq M$, and $r_{M+1} = 0$. Suppose that

$$a = q_1 b + r_1,$$
$$b = q_2 r_1 + r_2,$$
$$r_{i-1} = q_{i+1} r_i + r_{i+1}, \qquad i = 2, 3, \ldots, M.$$

Show that

$$\frac{a}{b} = q_1 + \cfrac{1}{q_2 + \cfrac{1}{q_3 \ddots + \cfrac{1}{q_{M+1}}}}.$$

5. Write a computer program that computes the GCD of two positive integers.

6. Let $a, b, m \in \mathbb{Z}$ with $a \neq 0$, $b \neq 0$ satisfy $\mathrm{GCD}(a, b) = 1$. Prove that if $a|m$ and $b|m$, then $ab|m$.

7. Let a and b be nonzero integers such that $\mathrm{GCD}(a, b) = 1$. Show that $\mathrm{GCD}(m, ab) = \mathrm{GCD}(m, a)\mathrm{GCD}(m, b)$ for all $m \in \mathbb{Z}$.

8. Let D_n be the set of positive divisors of a positive integer n. Given two positive integers m and n, define a function $\psi\colon D_m \times D_n \to D_{mn}$ by the formula $\psi(a, b) = ab$. Prove that if $\mathrm{GCD}(m, n) = 1$, then ψ is a bijection. What is the inverse?

9. Define $d(n)$ to be the number of positive divisors of a positive integer n. Prove that $d(mn) = d(m)d(n)$ if $\mathrm{GCD}(m, n) = 1$. Compute $d(p^a q^b r^c)$ where p, q, and r are distinct primes.

10. Let $\sigma(n)$ be the sum of the positive divisors of a positive integer n. Prove that $\sigma(mn) = \sigma(m)\sigma(n)$ if $\mathrm{GCD}(m, n) = 1$. Compute $\sigma(p^a q^b r^c)$ where p, q, and r are distinct primes.

11. Show that every rational number can be written in one and only one way as m/n with integers m, n such that $\mathrm{GCD}(m, n) = 1$ and $n > 0$.

12. A positive integer is called a *perfect number* iff it is equal to the sum of all its positive divisors other than itself.

i. Prove that $2^{p-1}(2^p - 1)$ is a perfect number if $2^p - 1$ is prime. Find the first four perfect numbers that arise in this way.

ii. Prove that every even perfect number is of the form $2^{p-1}(2^p - 1)$ where both p and $2^p - 1$ are prime. (*Hint:* Let n be even perfect. Write $n = 2^{p-1}s$ with s odd.) It is not known whether there are any odd perfect numbers.

13. Let $m, a_1, a_2, \ldots, a_N$ be integers such that $\mathrm{GCD}(m, a_i) = 1$ for all i. Prove that $\mathrm{GCD}(m, \prod_{i=1}^{N} a_i) = 1$.

14. We say that $c \neq 0 \in \mathbb{Z}$ is a *least common multiple* of two nonzero integers a, b iff c is divisible by both a and b and itself divides any other integer with the same property. Prove that $ab/\mathrm{GCD}(a, b)$ is a least common multiple of a and b.

15. For a positive integer n and prime number p, define $\mathrm{ord}_p(n)$ to be that integer $t \geq 0$ such that $p^t | n$ but $p^{t+1} \nmid n$. Write n as a product of primes, say $n = p_1 p_2 \cdots p_r$. Using only Lemma 3.4, prove that $\mathrm{ord}_p(n)$ equals the number of times p occurs among the factors $p_1, p_2, \ldots, p_r$. Hence prove Theorem 3.5.

16. Let $n = \prod p_i^{n_i}$ where the p_i are distinct prime numbers and the n_i are nonnegative integers. Prove that every nonzero integer a that divides n has a factorization $a = \pm \prod p_i^{a_i}$ where $0 \leq a_i \leq n_i$ for all i.

17. Prove that $\log_{10} 2$ is irrational.

18. Let $a, b, m \in \mathbb{Z}$ with $m \neq 0$ and $\mathrm{GCD}(a, b) = 1$.

i. Prove that if $ab = m^3$, then a and b are cubes of integers.

ii. What conclusion about a and b can be drawn if $ab = m^2$?

19. Let $a \neq 0 \in \mathbb{Q}$. Prove that a can be uniquely expressed as $a = m^2 n$ where $m \in \mathbb{Q}$ and n is a *squarefree integer* (i.e., n is not divisible by the square of any prime).

4. $a_1X_1 + a_2X_2 + \cdots + a_rX_r = n$

Proposition 4.1. Let $a, b, n \in \mathbb{Z}$ with a and b not both zero. The following two assertions are equivalent.

1. $aX + bY = n$ has a solution in integers X, Y.
2. $\mathrm{GCD}(a, b) | n$.

Proof. $1 \Rightarrow 2$. Suppose that $X, Y \in \mathbb{Z}$ and that $aX + bY = n$. Since $\mathrm{GCD}(a, b)$ divides both a and b, it must divide both aX and bY and hence also their sum, which is n.

$2 \Rightarrow 1$. Suppose that $n = t \cdot \mathrm{GCD}(a, b)$, where $t \in \mathbb{Z}$. By Proposition 3.3 there exist integers x, y such that $ax + by = \mathrm{GCD}(a, b)$. Then $aX + bY = n$ with $X = xt$ and $Y = yt$. ∎

Note that the proof of the implication $2 \Rightarrow 1$ in Proposition 4.1 gives a method of finding an integral solution of the equation $aX + bY = n$ if it has one. Proposition 4.2 will tell how to find all such solutions if just one is known.

Proposition 4.2. Let $a, b, n \in \mathbb{Z}$ with a and b not both zero. Let $x_0, y_0 \in \mathbb{Z}$ and suppose that $X = x_0$, $Y = y_0$ is a solution of the equation $aX + bY = n$. Then all solutions in integers X, Y of the equation $aX + bY = n$ are obtained through the formulas $X = x_0 + (b/d)t$, $Y = y_0 - (a/d)t$, where $d = \mathrm{GCD}(a, b)$ and $t \in \mathbb{Z}$.

Proof. First note that all X, Y obtained through the stated formulas actually are integer solutions of the equation $aX + bY = n$. It remains to prove that there are no other solutions.

Taking the difference of $aX + bY = n$ and $ax_0 + by_0 = n$ and then dividing by d yields the equation

$$\frac{a}{d}(X - x_0) + \frac{b}{d}(Y - y_0) = 0. \qquad (*)$$

Suppose that $b \neq 0$. Then $b/d \mid (a/d)(X - x_0)$. By Lemma 3.7 we know that a/d and b/d are relatively prime. It then follows from Lemma 3.6 that $b/d \mid X - x_0$, so there exists $t \in \mathbb{Z}$ such that $X - x_0 = (b/d)t$. Substituting back into $(*)$ shows that $Y - y_0 = -(a/d)t$. Finally, if $b = 0$, the equation is $aX = n$, which has general solution $X = n/a = x_0$ and Y an arbitrary integer. The proposition holds because in this case $a/d = \pm 1$. ∎

We next extend the notion of greatest common divisor, as preparation for a study of the equation $a_1X_1 + a_2X_2 + \cdots + a_rX_r = n$.

Definition. An integer c is a *greatest common divisor* for a set $S \subset \mathbb{Z}$ iff c divides every element of S and is itself divisible by all other integers with the same property.

The existence of greatest common divisors is taken care of by the following lemma, which in fact shows how to compute them.

Lemma 4.3. Let $a_1, a_2, \ldots \in \mathbb{Z}$ with $a_1 \neq 0$. Let $c_1 = a_1$ and let $c_i = \mathrm{GCD}(c_{i-1}, a_i)$ for $i \geq 2$. Then c_r is a greatest common divisor for $\{a_1, a_2, \ldots, a_r\}$ for all $r \geq 1$. Moreover, there exist integers x_i such that $\sum_{i=1}^{r} a_i x_i = c_r$.

Proof. By induction on r. There is nothing to prove if $r = 1$, so suppose $n > 1$ and that the lemma has been proved for $r = n - 1$. The inductive hypothesis asserts that c_{n-1} is a GCD for $\{a_1, \ldots, a_{n-1}\}$. A common divisor of $a_1, a_2, \ldots, a_{n-1}, a_n$ divides each of $a_1, a_2, \ldots, a_{n-1}$ and must therefore divide c_{n-1}; because it also divides a_n, it must divide $c_n = \mathrm{GCD}(c_{n-1}, a_n)$ as well. On the other hand, c_n divides both c_{n-1} and a_n and hence divides each of $a_1, a_2, \ldots, a_n$. Therefore c_n is a GCD for $\{a_1, a_2, \ldots, a_n\}$.

Finally, by the inductive hypothesis there exist $x_i' \in \mathbb{Z}$ such that $\sum_{i=1}^{n-1} a_i x_i' = c_{n-1}$. By Proposition 3.3 there exist $x, y \in \mathbb{Z}$ such that $c_{n-1}x + a_n y = c_n$. We calculate

$$c_n = \left(\sum_{i=1}^{n-1} a_i x_i' \right) x + a_n y$$

$$= \sum_{i=1}^{n-1} a_i (x_i' x) + a_n y.$$

Thus $c_n = \sum_{i=1}^{n} a_i x_i$, with $x_i = x_i' x$ for $i < n$ and $x_n = y$. ■

A finite set of integers not all zero has exactly two GCDs differing only in sign, as in the case of GCDs for pairs of integers. We will write $\mathrm{GCD}(a_1, \ldots, a_n)$ for the *positive* greatest common divisor.

Theorem 4.4. Let $a_1, a_2, \ldots, a_r, n \in \mathbb{Z}$ with the a_i not all zero. The following two statements are equivalent:

1. $a_1X_1 + a_2X_2 + \cdots + a_rX_r = n$ *has a solution in integers* X_i.
2. $\mathrm{GCD}(a_1, a_2, \ldots, a_r) | n$.

Proof. Formally identical to the proof of Proposition 4.1. ■

We next ask for *all* integral solutions of the equation $\sum a_i X_i = n$. The key lemma is as follows.

Lemma 4.5. Let $\underline{a} = (a_1, \ldots, a_r) \neq 0 \in \mathbb{Z}^r$ and let $d = \mathrm{GCD}(a_1, a_2, \ldots, a_r)$. Then there exists an $r \times r$ matrix C with integer entries and determinant equal to ± 1 such that

$$\underline{a}C = (0, 0, \ldots, 0, d).$$

Proof. The proof is by induction on r, the length of the vector $\underline{a}$.

If $r = 1$, then $d = |a_1|$, so we may take $C = (\operatorname{sign}(a_1))$.

Now let $r \geq 2$, and suppose that Lemma 4.5 has been proved for all vectors of length less than r. If $a_1 = a_2 = \cdots = a_{r-1} = 0$, then we may take

$$C = \begin{pmatrix} I_{r-1} & \vdots & 0 \\ \cdots & \vdots & \cdots \\ 0 & \cdot & \operatorname{sign}(a_r) \end{pmatrix},$$

where I_{r-1} is the $(r-1) \times (r-1)$ identity matrix and the two zeroes represent matrices all of whose entries are zero. So suppose that $(a_1, \ldots, a_{r-1}) \neq 0 \in Z^{r-1}$. Let $c = \operatorname{GCD}(a_1, \ldots, a_{r-1})$. By the inductive hypothesis there is an $(r-1) \times (r-1)$ matrix A with integer entries and determinant equal to ± 1 such that $(a_1, \ldots, a_{r-1})A = (0, \ldots, 0, c) \in \mathbb{Z}^{r-1}$. Let A^* be the $r \times r$ matrix

$$A^* = \begin{pmatrix} A & \vdots & 0 \\ \cdots & \vdots & \cdots \\ 0 & \vdots & 1 \end{pmatrix}.$$

We have $\underline{a}A^* = (0, \ldots, 0, c, a_r)$. By Lemma 4.3, $d = \operatorname{GCD}(c, a_r)$. Therefore by Proposition 3.3 there exist integers x, y such that $cx + a_r y = d$. The 2×2 integer matrix

$$B = \begin{pmatrix} \dfrac{a_r}{d} & x \\ -\dfrac{c}{d} & y \end{pmatrix},$$

which has determinant equal to $+1$, satisfies the equation $(c, a_r)B = (0, d)$. Hence the $r \times r$ integer matrix

$$C = A^* \begin{pmatrix} I_{r-2} & \vdots & 0 \\ \cdots & \vdots & \cdots \\ 0 & \vdots & B \end{pmatrix}$$

has determinant equal to ± 1 and satisfies the equation $\underline{a}C = (0, \ldots, 0, d)$ as required. ■

Note that the proof of existence of C in Lemma 4.5 is quite constructive; it shows how to calculate an explicit matrix C for any $\underline{a}$. This calculation amounts to the complete resolution of the linear Diophantine equation $a_1X_1 + a_2X_2 + \cdots + a_rX_r = n$, as is shown by our next theorem.

Theorem 4.6. Let $\underline{a}$, d, and C be as in Lemma 4.5. For $i = 1, 2, \ldots, r$, let $\underline{c}_i$ denote the ith column of the matrix C. Let $N \in \mathbb{Z}$.

i. $\underline{x} = \begin{pmatrix} x_1 \\ \vdots \\ x_r \end{pmatrix} \in \mathbb{Z}^r$ solves the equation $\sum_{i=1}^r a_i x_i = Nd$ if and only if there exists integers m_i such that $\underline{x} = \sum_{i=1}^{r-1} m_i \underline{c}_i + N\underline{c}_r$.

ii. The set of integer solutions of the equation $\sum_{i=1}^r a_i X_i = 0$ is a subgroup of $\mathbb{Z}^r$ that is isomorphic to $\mathbb{Z}^{r-1}$.

Proof. i. Let $\underline{y} = \underline{x} - N\underline{c}_r$. Then

$$\sum_{i=1}^r a_i x_i = Nd \Leftrightarrow \underline{a}\,\underline{y} = 0$$

$$\Leftrightarrow \underline{a}C\left(C^{-1}\underline{y}\right) = (0 \cdots 0d)C^{-1}\underline{y} = 0$$

$$\Leftrightarrow C^{-1}\underline{y} = \underline{m} = \begin{pmatrix} m_1 \\ \vdots \\ m_{r-1} \\ 0 \end{pmatrix} \quad \text{for some integers } m_1, m_2, \ldots, m_{r-1}$$

$$\Leftrightarrow \underline{y} = C\underline{m} = m_1\underline{c}_1 + m_2\underline{c}_2 + \cdots + m_{r-1}\underline{c}_{r-1}.$$

ii. For $\underline{m} = (m_1, \ldots, m_{r-1}) \in \mathbb{Z}^{r-1}$, define $\phi(\underline{m}) = \sum_{i=1}^{r-1} m_i \underline{c}_i$. By part i, ϕ is a surjective homomorphism from $\mathbb{Z}^{r-1}$ onto the group of integer solutions of the equation $\sum_{i=1}^r a_i X_i = 0$. Because C has nonzero determinant, its columns $\underline{c}_i$ are linearly independent. This implies that ϕ is injective. Hence ϕ is an isomorphism. ■

As an example, consider the equation $5X + 7Y + 11Z = 2$. The matrix

$$C = \begin{pmatrix} 7 & -44 & -4 \\ -5 & 33 & 3 \\ 0 & -1 & 0 \end{pmatrix}$$

is as demanded in Lemma 4.5 where we take $a = (5, 7, 11)$. The complete integral solution of the equation can therefore be given as

$$X = 7m - 44n - 8,$$
$$Y = -5m + 33n + 6,$$
$$Z = -n, \quad \text{where } m, n \in \mathbb{Z}.$$

The theory developed in this section can be extended to include systems of s linear Diophantine equations in r variables. The extension, called the theory of elementary divisors, is perhaps best presented as a study of canonical forms for group homomorphisms from $\mathbb{Z}^r$ to $\mathbb{Z}^s$. It is of great importance but lies beyond the scope of this book.

Exercises

1. Let $a = 1587645$ and $b = 6755$. Compute $d = \text{GCD}(a, b)$. Find all integers X, Y such that $aX + bY = d$.

2. Let a, b be positive integers with $\text{GCD}(a, b) = 1$. Show that there exist integers $x, y \geq 0$ such that $ax + by = n$ for every integer $n > ab - a - b$, but that there do not exist such integers for $n = ab - a - b$.

3. Find all integer solutions of the equation $2X^2 - XY - 3Y^2 = 8$.

4. Find all integer solutions:

i. $2X + 3Y + 5Z = 1$;

ii. $6X + 15Y + 35Z = 1$.

5. Write a computer program that accepts as input a vector $\underline{a} = (a_1, \ldots, a_r) \neq 0 \in \mathbb{Z}^r$ and prints out a matrix C as in Lemma 4.5.

6. Deduce Theorem 4.4 from Lemma 4.5.

7. Let $\phi\colon \mathbb{Z}^r \to \mathbb{Z}$ be a nonzero homomorphism of abelian groups. Prove that $\ker(\phi) \simeq \mathbb{Z}^{r-1}$ and that $\text{im}(\phi) \simeq \mathbb{Z}$.

8. Let a, b, c be nonzero integers that are pairwise relatively prime. Show that every integer solution (X, Y, Z) of the equation $bcX + acY + abZ = 0$ equals $m(a, -b, 0) + n(0, b, -c)$ for some integers m, n.

9. i. Let $m_1, m_2, \ldots, m_r$ be nonzero integers that are pairwise relatively prime. Let $a \in \mathbb{Z}$. Show that there exist $a_1, a_2, \ldots, a_r \in \mathbb{Z}$ such that

$$\frac{a}{m_1 m_2 \cdots m_r} = \frac{a_1}{m_1} + \frac{a_2}{m_2} + \cdots + \frac{a_r}{m_r}.$$

ii. Find integers a, b, c such that

$$\frac{1}{8 \cdot 27 \cdot 125} = \frac{a}{8} + \frac{b}{27} + \frac{c}{125}.$$

10. Let a, b, c be positive integers with $\mathrm{GCD}(a, b, c) = d$. Show that there exists $N > 0$ such that $aX + bY + cZ = nd$ has a solution in *positive* integers X, Y, Z for every integer $n > N$.

11. Show that the number of solutions of $X + 2Y + 3Z = n$ with integers $X, Y, Z \geq 0$ equals the coefficient of x^n in the power series expansion of $1/((1 - x)(1 - x^2)(1 - x^3))$.

12. Find all integer solutions of the system of two equations:

$$2X + 3Y + 5Z = 0$$
$$3X + 5Y + 7Z = 0.$$

5. The Distribution of the Primes

Euclid proved that there is an infinite number of primes. In 1737 Euler took the first step beyond Euclid by proving that the sum $\Sigma 1/p$ of the reciprocals of the primes p diverges. His result can perhaps be interpreted as meaning, for instance, that there are "more" primes than squares, because $\Sigma_{n=1}^{\infty} 1/n^2$ converges. With one brilliant stroke Euler brought analysis into number theory and initiated the quantitative study of the distribution of prime numbers.

For $x > 0$ let $\pi(x)$ equal the number of primes p with $p \leq x$. In the hands of Riemann and then Hadamard and de la Vallée Poussin, Euler's methods led to a proof of the great Prime Number Theorem that had been conjectured by Legendre and Gauss, namely that $\lim_{x \to \infty} \pi(x)/(x/\log x) = 1$. The Prime Number Theorem will not be demonstrated here. We will prove only Chebyshev's elementary theorem that $\pi(x)$ and $x/\log x$ are of the same order of magnitude. That will be enough to deduce that the proportion $\pi(x)/x$ of positive integers less than x that are prime tends to zero as $x \to \infty$.

In the rest of this section Σ_p and Π_p will denote the sum and product over all prime numbers p having the stated property.

Theorem 5.1. Euler. $\Sigma 1/p = \infty$.

We need some lemmas.

Lemma 5.2. Let

$$q(N) = \prod_{p \leq N} \frac{1}{1 - 1/p} \quad \text{for } N \geq 2.$$

Then $\lim_{N \to \infty} q(N) = \infty$.

Proof. We make use of the geometric series

$$\frac{1}{1 - 1/p} = \sum_{a=0}^{\infty} \frac{1}{p^a}.$$

Thus for example

$$\frac{1}{1 - 1/2} \frac{1}{1 - 1/3} = \sum_{a=0}^{\infty} \sum_{b=0}^{\infty} \frac{1}{2^a 3^b}.$$

(There is no repetition among the denominators $2^a 3^b$ in the sum because of the uniqueness of factorization, but we do not need this fact.)

Similarly, $q(N) = \Sigma_{n \in X}(1/n)$, where X is the set of positive integers that are products of primes less than or equal to N. In particular, X contains the numbers $1, 2, \ldots, N$. Therefore $q(N) > \Sigma_{n=1}^{N}(1/n)$. The lemma now follows because $\Sigma_{n=1}^{\infty}(1/n) = \infty$. ■

Corollary 5.3. There is an infinite number of primes.

Proof. If not, then $\lim_{N \to \infty} q(N)$ would be a finite product and would therefore be finite. ■

Lemma 5.4.

$$-\log(1 - x) = x + \frac{x^2}{2} + \frac{x^3}{3} + \cdots = \sum_{m=1}^{\infty} \frac{x^m}{m} \quad \text{for } |x| < 1.$$

Proof. The two sides of the equation are equal at $x = 0$. Moreover, the two sides have the same derivatives, since $1/(1 - x) = 1 + x + x^2 + x^3 + \cdots$. Therefore, by the mean value theorem, the two sides are equal. ■

Proof of Theorem 5.1.

$$\begin{aligned}
\log q(N) &= \sum_{p \le N} \log \frac{1}{1 - 1/p} \\
&= \sum_{p \le N} \sum_{m=1}^{\infty} \frac{1}{mp^m} \\
&= \sum_{p \le N} \frac{1}{p} + \sum_{p \le N} \sum_{m=2}^{\infty} \frac{1}{mp^m}.
\end{aligned}$$

From

$$\sum_{m=2}^{\infty} \frac{1}{mp^m} < \sum_{m=2}^{\infty} \frac{1}{p^m} = \frac{1}{p(p-1)} < \frac{1}{(p-1)^2}$$

we learn that

$$\sum_{p \le N} \sum_{m=2}^{\infty} \frac{1}{mp^m} < \sum_{n=1}^{\infty} \frac{1}{n^2} = A,$$

where A is a constant independent of N.

Therefore,

$$\sum_{p \le N} \frac{1}{p} > \log q(N) - A.$$

Theorem 5.1 now follows from Lemma 5.2. ∎

We conclude Section 5 with the proof of a slightly strengthened version of Theorem 1.2.

Theorem 5.5. Chebyshev.

$$\frac{\log 2}{4} \frac{x}{\log x} < \pi(x) < 6 \log 2 \frac{x}{\log x} \quad \text{for } x \ge 2.$$

(Note the approximations $(\log 2)/4 > 0.17$ and $6 \log 2 < 4.16$.)

Corollary 5.6. $\lim_{x \to \infty} \pi(x)/x = 0$.

The proof of Chebyshev's theorem is based on an analysis of the size and factorization of the binomial coefficients $\binom{2n}{n} = (2n)!/n!n!$. We take the two inequalities separately, beginning with the upper bound.

Lemma 5.7.

$$n^{\pi(2n) - \pi(n)} \le \prod_{n < p \le 2n} p \le \binom{2n}{n} < 2^{2n} \quad \text{for } n \ge 1.$$

Proof. The first inequality is obvious upon noting that $\pi(2n) - \pi(n)$ is the number of primes p with $n < p \le 2n$.

To prove the second, observe that $\prod_{n<p\le 2n} p$ divides the positive integer $\binom{2n}{n}$ because the given primes divide the numerator but not the denominator of $(2n)!/n!n!$.

Finally, $\binom{2n}{n}$ is but the middle term of the binomial expansion of $(1+1)^{2n}$. Since all the terms are positive, we conclude that $\binom{2n}{n} < 2^{2n}$. ■

Proof that $\pi(x) < 6\log 2(x/\log x)$. Let $n = 2^{k-1}$ in Lemma 5.7. Comparing exponents, we find that $(k-1)(\pi(2^k) - \pi(2^{k-1})) < 2^k$. Therefore, $k\pi(2^k) < (k-1)\pi(2^{k-1}) + 3\cdot 2^{k-1}$ for $k \ge 1$, where we have made use of the inequality $\pi(2^k) \le 2^{k-1}$, whose proof is left as an exercise. An induction on k establishes the basic inequality:

$$\pi(2^k) < 3\cdot\frac{2^k}{k} \quad \text{for } k \ge 1.$$

Now let x satisfy $2^k \le x < 2^{k+1}$. We compute

$$\pi(x) \le \pi(2^{k+1}) < 3\cdot\frac{2^{k+1}}{k+1} = 6\log 2\frac{2^k}{\log 2^{k+1}} < 6\log 2\frac{x}{\log x}. \qquad ■$$

For any real number x we will write $[x]$ for the largest integer that is less than or equal to x. That is, $[x] \in \mathbb{Z}$ and $[x] \le x < [x]+1$.

Lemma 5.8. Let p be a prime number, let n be a positive integer, and let p^t be the highest power of p that divides $n!$. Then $t = \sum_{j=1}^{\infty}[n/p^j]$.

Proof. From the inequalities $[n/p^j]p^j \le n < ([n/p^j]+1)p^j$, we find that $[n/p^j]$ is the number of those integers $1, 2, \ldots, n$ that are divisible by p^j. Hence $[n/p^j] - [n/p^{j+1}]$ is the number of integers $m \in \{1, 2, \ldots, n\}$ that can be written $m = p^j s$ where $p \nmid s$. Therefore, $t = \sum_{j=1}^{\infty} j([n/p^j] - [n/p^{j+1}])$, which telescopes into the sum $\sum_{j=1}^{\infty}[n/p^j]$ as claimed. ■

Lemma 5.9. Let p be prime, let $n \ge 1$, and let p^r be the highest power of p that divides $\binom{2n}{n} = (2n)!/n!n!$. Then $p^r \le 2n$.

Proof. Let $s \in \mathbb{Z}$ be such that $p^s \le 2n < p^{s+1}$. By Lemma 5.8,

$$r = \sum_{j=1}^{\infty}[2n/p^j] - 2[n/p^j]$$

$$= \sum_{j=1}^{s}[2n/p^j] - 2[n/p^j].$$

Since $[2x] - 2[x] = 0$ or 1 for all real numbers x, it follows that $r \leq s$. Hence $p^r \leq p^s \leq 2n$. ■

Proof that $\pi(x) > \frac{1}{4}(\log 2)(x / \log x)$. Note first that $2^n \leq \binom{2n}{n} \leq (2n)^{\pi(2n)}$ for $n \geq 1$. To see the first of these inequalities, write

$$\binom{2n}{n} = \frac{2n}{n} \cdot \frac{2n-1}{n-1} \cdots \frac{n+1}{1},$$

a product of n factors every one of which is at least 2. For the second, write

$$\binom{2n}{n} = \prod_{p \leq 2n} p^{r_p},$$

a product of $\pi(2n)$ prime powers that, by Lemma 5.9, are each at most $2n$.

Let $n = 2^{k-1}$ in the above inequality. Comparing exponents, we establish the basic inequality:

$$\pi(2^k) \geq 2^k/2k \quad \text{for } k \geq 1.$$

Now let x satisfy $2^k \leq x < 2^{k+1}$. We compute

$$\pi(x) \geq \pi(2^k) \geq \frac{2^k}{2k} = \frac{\log 2}{4} \frac{2^{k+1}}{\log 2^k} > \frac{\log 2}{4} \frac{x}{\log x}.$$ ■

Exercises

1. Prove that $\pi(n) \leq n/2$ for every positive integer n except for $n = 3, 5, 7$.

2. Show that $\pi(n) \leq n/3$ for all integers $n \geq 33$.

3. Prove that $[2x] - 2[x] = 0$ or 1 for all real numbers x.

4. Let m, n be positive integers. Show that $(2m)!(2n)!/(m!n!(m+n)!)$ is also an integer. What does it count?

5. For $x > 0$ let $a(x)$ equal the number of square integers n^2 with $n^2 \leq x$. Prove that $\lim_{x \to \infty} \pi(x)/a(x) = \infty$.

6. **i.** Prove that there exists $c > 0$ such that $p_n > cn \log n$ for $n \geq 1$, where p_n is the nth prime number.

ii. Show that there exists $A > 0$ such that $p_n < An^2$ for $n \geq 1$.

Hence prove that there exists $C > 0$ such that $p_n < Cn \log n$ for $n \geq 2$.

iii. Show that p_{n+1}/p_n is a bounded function of n.

7. **i.** Show that there exists $A > 0$ such that $|\log n! - n\Sigma_{p \le n}(\log p/p)| < An$ for all integers $n \ge 1$. (*Suggestion:* Start with Lemma 5.8. Use Theorem 1.2.)

ii. Show that $n \log n \ge \log n! > n \log n - n$ for $n \ge 1$.

iii. Prove that there exists $C > 0$ such that $|\Sigma_{p \le x}(\log p/p) - \log x| < C$ for all real numbers $x \ge 1$.

8. **i.** Define $B(x)$ for $x \ge 1$ by $B(x) = \Sigma_{p \le x}(\log p/p)$. Verify the following equalities for integers $N \ge 2$:

$$\sum_{p \le N} \frac{1}{p} = \sum_{i=2}^{N-1} B(i)\left(\frac{1}{\log(i)} - \frac{1}{\log(i+1)}\right) + B(N)\frac{1}{\log N}$$

$$= \int_2^N \frac{B(x)}{x(\log x)^2}\,dx + \frac{B(N)}{\log N}.$$

ii. Prove that $\lim_{N \to \infty}(\Sigma_{p \le N}(1/p) - \log\log N)$ exists and is finite.

9. Prove that there exist $C > c > 0$ such that

$$c \log N < \prod_{p \le N} \frac{1}{1 - 1/p} < C \log N \quad \text{for } N \ge 2.$$

10. In this exercise we write p for all primes, q for all primes of the form $4k + 1$, and r for primes of the form $4k + 3$. We set $\epsilon(2) = 0$, $\epsilon(q) = 1$, and $\epsilon(r) = -1$, so that $\epsilon(p) = (-1)^{(p-1)/2}$ for odd primes p.

i. Working purely formally (like Euler), argue that

$$\lim_{N \to \infty} \prod_{p \le N} \frac{1}{1 - \epsilon(p)/p} = \frac{1}{1} - \frac{1}{3} + \frac{1}{5} - \frac{1}{7} + \frac{1}{9} + \cdots$$

$$= \sum_{\substack{n=1 \\ n \text{ odd}}}^{\infty} \frac{(-1)^{(n-1)/2}}{n}.$$

Show that the alternating series converges to a *positive* real number. (It converges to $\pi/4$ as can be proved by evaluating $\pi\cot(\pi t) = (1/t) + \Sigma_{m=1}^{\infty}(1/(m+t) - 1/(m-t))$ at $t = \frac{1}{4}$.)

ii. Assuming the equality of i, show that $\Sigma\epsilon(p)/p$ converges, where the primes p are taken in their natural order. Deduce that $\Sigma 1/q = \infty$ and that $\Sigma 1/r = \infty$ and hence that there are infinitely many primes of each of the two forms $4k + 1$ and $4k + 3$.

iii. The statements of ii can all be proven. Fill in the details of the following sketch. The series $L(x) = \Sigma_{n=1,\ n\ \text{odd}}^{\infty}(-1)^{(n-1)/2}/n^x$ converges uniformly in the region $x \geq 1$ and absolutely for $x > 1$. Note that $L(x) > 0$ for $x \geq 1$. For $N \geq 2$ and $x \geq 1$ let $L_N(x) = \Pi_{p \leq N}(1 - \epsilon(p)/p^x)^{-1}$. Then $\lim_{N\to\infty} L_N(x) = L(x)$ for $x > 1$. (The assertion of i is that $\lim_{x\downarrow 1}\lim_{N\to\infty} L_N(x) = \lim_{N\to\infty}\lim_{x\downarrow 1} L_N(x)$, a calculation that it is the object of these remarks to circumvent.) Write $\log L_N(x) = \Sigma_{p \leq N}\epsilon(p)/p^x + A_N(x)$, where easy estimates show that $\lim_{N\to\infty} A_N(x)$ is continuous for $x \geq 1$. Because $L(1) > 0$, we find that $\lim_{x\downarrow 1}\Sigma\epsilon(p)/p^x$ exists and is finite. We know from Theorem 5.1 that $\lim_{x\downarrow 1}\Sigma 1/p^x = \infty$. Adding and subtracting the last two limits shows that $\lim_{x\downarrow 1}\Sigma 1/q^x = \lim_{x\downarrow 1}\Sigma 1/r^x = \infty$, whence $\Sigma 1/q = \Sigma 1/r = \infty$.

With more work (see Landau, *Handbuch der Lehre von der Verteilung der Primzahlen*, Section 109) it can be shown that $\lim_{x\downarrow 1}\Sigma\epsilon(p)/p^x = \Sigma\epsilon(p)/p$, from which it follows that $\lim_{N\to\infty} L_N(x)$ exists and is continuous at $x = 1$. This establishes the amazing formula of i discovered by Euler:

$$\frac{\pi}{4} = \frac{3}{4}\,\frac{5}{4}\,\frac{7}{8}\,\frac{11}{12}\,\frac{13}{12}\,\frac{17}{16}\cdots,$$

where the numerators are the odd primes and the denominators are the adjacent multiples of 4.

iv. The sketch in iii touches all the essential points of a general proof of Dirichlet's theorem on primes in arithmetic progressions. The deepest and most interesting point turns out to be the assertion that $L(1) \neq 0$ for appropriate functions L, which Dirichlet proved by explicit evaluation of $L(1)$.

CHAPTER 2
Sums of Two Squares

1. Introduction

In this chapter we discuss several ideas linked by their relationship to the following famous theorem of Fermat.

Theorem 1.1. $X^2 + Y^2 = p$ has a solution in integers X, Y for every prime number p that is congruent to 1 mod 4.

This is given force by:

Theorem 1.2. There are infinitely many primes congruent to 1 mod 4.

We can use Theorem 1.1 as the basis for a complete determination of the integers that are sums of two (integer) squares.

Theorem 1.3. The following are equivalent for a positive integer n.

1. $X^2 + Y^2 = n$ has a solution in integers X, Y.
2. All primes of the form $4k + 3$ that divide n appear in the prime factorization of n with even exponent. In other words,

$$n = 2^a p_1^{b_1} \cdots p_r^{b_r} q_1^{c_1} \cdots q_1^{c_s}$$

with p_i distinct primes $\equiv 1 \pmod 4$, q_i distinct primes $\equiv 3 \pmod 4$, and c_i even integers.

We present five proofs of Theorem 1.1, based, respectively, on:

- unique factorization in the Gaussian integers — Section 4
- rational approximation of real numbers — Section 5
- Minkowski's theorem — Section 6
- method of descent — Section 7
- reduction of positive definite quadratic forms — Section 8

The starting point for all these proofs is the existence half of the following lemma, which will be proved twice in Section 2.

Lemma 1.4 (main lemma). $X^2 \equiv -1 \pmod{p}$ has a solution for an odd prime p if and only if p is congruent to 1 mod 4.

This lemma says that $X^2 + Y^2 =$ (multiple of p) has a nonzero solution if p is congruent to 1 mod 4. Naturally we will want to show that the multiple of p can be taken to be p itself.

We prove Theorem 1.2 and the implication Theorem 1.1 $\Rightarrow$ Theorem 1.3 in Section 3.

There is a proof that every positive integer is a sum of four squares in Section 7.

Exercise

1. Let $n \in \mathbb{Z}$. Deduce from Theorem 1.3 that if there exist $p, q \in \mathbb{Q}$ such that $p^2 + q^2 = n$, then there exist $a, b \in \mathbb{Z}$ such that $a^2 + b^2 = n$.

2. Integers mod *m*

Many proofs in number theory can be viewed as extended sequences of deductions about the divisibility properties of various integers. The deductions can be greatly facilitated by the simple device of expressing the divisibility relations as equations that can be transformed in accordance with the ordinary rules of algebra. Classically one works with congruence equations between integers. In contemporary mathematics, congruences take the form of ordinary equations in the rings that are quotient rings of the ring $\mathbb{Z}$ of integers. It is the purpose of Section 2 to introduce this theory.

Definition. The *ring* $\mathbb{Z}/m$ *of integers* mod m is defined for integers $m \neq 0$ to be the quotient ring $\mathbb{Z}/m\mathbb{Z}$.

The *group* U_m *of invertible integers* mod m is the group of units of the ring $\mathbb{Z}/m$.

For $x \in \mathbb{Z}$, we write $\bar{x} \in \mathbb{Z}/m$ for the image of x under the canonical homomorphism $\mathbb{Z} \to \mathbb{Z}/m$.

For $x, y \in \mathbb{Z}$, we write $x \equiv y \pmod{m}$ (read x congruent to y mod m) iff $\bar{x} = \bar{y} \in \mathbb{Z}/m$. (Equivalently, iff $m|(x - y)$.)

Clearly $|\mathbb{Z}/m| = |m|$. (We write $|G|$ for the order of a finite group G.)

We investigate U_m.

Proposition 2.1. Let $x \in \mathbb{Z}$. Then $\bar{x} \in U_m$ if and only if $\mathrm{GCD}(x, m) = 1$.

Proof. ($\Rightarrow$) Let $\bar{x} \in U_m$. Then there exists $y \in \mathbb{Z}$ with $\overline{xy} = \bar{1}$; that is, $xy \equiv 1 \pmod{m}$; so $xy - tm = 1$ for some $t \in \mathbb{Z}$. From this we see that x and m can have no common prime factors.

($\Leftarrow$) If $\mathrm{GCD}(x, m) = 1$, then $xy + tm = 1$ is solvable in integers y and t. But then $\overline{xy} = \bar{1}$. Thus $\bar{x}$ is invertible. ■

Corollary 2.2. $|U_p| = p - 1$ if p is prime. Thus $\mathbb{Z}/p$ is a field.

Corollary 2.3. Fermat's Little Theorem. Let p be prime and $\mathrm{GCD}(a, p) = 1$. Then $a^{p-1} \equiv 1 \pmod{p}$. (Equivalently, $a^{p-1} = \bar{1}$ for every $a \in U_p$.)

Proof 1. The order l of $a \in U_p$ must divide the order of U_p. That is, l divides $p - 1$. Hence $(p - 1)/l \in \mathbb{Z}$. We calculate $a^{p-1} = (a^l)^{(p-1)/l} = \bar{1}$.

Proof 2. Let $\mathrm{GCD}(a, p) = 1$. Write down two sequences of integers.

$$\begin{array}{ccccc} 1 & 2 & 3 & \cdots & p-1 \\ a & 2a & 3a & \cdots & (p-1)a \end{array}$$

The first sequence is a complete set of representatives of the nonzero congruence classes mod p.

We claim that the second sequence is too. Each element of the second sequence is nonzero mod p, being the product of integers that are nonzero mod p, and no two integers in the second sequence are congruent mod p, since $p \nmid a$ (just see whether p divides the difference of two of them!). So the second sequence is a subset of a set of representatives. Since it has $p - 1$ elements, it must in fact be a complete set of representatives of the nonzero congruence classes mod p.

Thus the elements of the second sequence, after rearranging, must be congruent to the elements in the first.

$$1 \cdot 2 \cdot 3 \cdots (p-1) \equiv a \cdot 2a \cdot 3a \cdots (p-1)a \pmod{p},$$

$$(p-1)! \equiv (p-1)! a^{p-1} \pmod{p}.$$

Since $(p-1)!$ is invertible mod p, we conclude $1 \equiv a^{p-1} \pmod{p}$. ∎

Lemma 2.4. Wilson's Theorem. $(p-1)! \equiv -1 \pmod{p}$ if p is prime.

Proof. We ask which elements in the following sequence equal their own inverses mod p.

$$1 \quad 2 \quad 3 \quad \cdots \quad p-1.$$

Answer: Only 1 and $p-1$. For if $\bar{x} \cdot \bar{x} = \bar{1}$, that is, if $x^2 \equiv 1 \pmod{p}$, then $p | x^2 - 1$. But $p | (x-1)(x+1)$ means that either $p | (x-1)$, giving $x \equiv 1 \pmod{p}$, or $p | (x+1)$, giving $x \equiv -1 \pmod{p}$.

So we compute $(p-1)!$ by pairing inverses thus:

$$\begin{aligned} (p-1)! &\equiv 1 \cdot (p-1) \cdot (2 \cdot 2^{-1})(3 \cdot 3^{-1}) \cdots \\ &\equiv 1 \cdot (p-1) \equiv -1 \pmod{p}. \end{aligned}$$ ∎

We come to a refined version of Lemma 1.4 of the Introduction, the principal lemma in Chapter 2.

Proposition 2.5. Let p be an odd prime.

Then $X^2 \equiv -1 \pmod{p}$ is solvable if and only if $p \equiv 1 \pmod{4}$, in which case $X = \pm((p-1)/2)!$ are the only two solutions mod p.

Proof. ($\Rightarrow$) Suppose that $x^2 \equiv -1 \pmod{p}$. Then

$$(x^2)^{(p-1)/2} \equiv x^{p-1} \equiv 1 \pmod{p},$$

$$|||$$

$$(-1)^{(p-1)/2} \equiv \begin{cases} 1 & \text{if } p \equiv 1 \pmod{4} \\ -1 & \text{if } p \equiv 3 \pmod{4} \end{cases}.$$

So we must have $p \equiv 1 \pmod 4$.

$$(\Leftarrow)\quad \text{(Lagrange)} -1 \equiv (p-1)! = \left(1 \cdot 2 \cdots \frac{p-1}{2}\right) \cdot \left(\frac{p+1}{2} \cdots p-1\right)$$

$$= \left(1 \cdot 2 \cdots \frac{p-1}{2}\right) \cdot \left(p - \frac{p-1}{2} \cdots p-1\right)$$

$$\equiv \left(\frac{p-1}{2}\right)! \cdot (-1)^{(p-1)/2}\left(\frac{p-1}{2}\right)!$$

$$\equiv (-1)^{(p-1)/2}\left(\frac{p-1}{2}\right)!^2 \pmod p.$$

Thus

$$\left(\frac{p-1}{2}\right)!^2 \equiv \begin{cases} -1 & \text{if } p \equiv 1 \pmod 4 \\ 1 & \text{if } p \equiv 3 \pmod 4 \end{cases}.$$

For the final assertion, we note that $a^2 \equiv b^2 \pmod p$ implies that $a^2 - b^2 \equiv 0 \pmod p$. Thus $p \mid (a+b)(a-b)$. So either $p \mid (a+b)$, giving $a \equiv -b \pmod p$, or $p \mid (a-b)$, giving $a \equiv b \pmod p$. Apply this to $b = ((p-1)/2)!$ ∎

Notice that in the proofs of both Lemma 2.4 and Proposition 2.5 we needed to know that $X^2 \equiv c \pmod p$ has at most two solutions mod p. (In Lemma 2.4, $c = 1$. In Proposition 2.5, $c = b^2 = -1$.) Rephrased, we needed to know that $X^2 - c$ has at most two roots in $\mathbb{Z}/p$. We prove more in the next lemma, where we have written $\mathbb{Z}/p[X]$ for the ring of polynomials with coefficients in $\mathbb{Z}/p$. The result depends crucially on the fact that $\mathbb{Z}/p$ is a field.

Lemma 2.6. Let p be prime and let $f(X) \in \mathbb{Z}/p[X]$ be a nonzero polynomial of degree s. Then $f(X)$ has at most s roots in $\mathbb{Z}/p$.

Proof. Argument by induction. If a is a root of $f(X)$, then $X - a$ divides $f(X)$ (division algorithm). So $f(X) = (X-a)g(X)$ where $\deg g(X) = s - 1$. Because $\mathbb{Z}/p$ is a field, roots of $f(X)$ distinct from a must be roots of $g(X)$. The inductive hypothesis shows that there can be at most $s-1$ of them. ∎

Polynomial arithmetic suggests an alternative proof of Wilson's theorem as follows.

Proof 2 of Lemma 2.4. By Fermat's Little Theorem, Corollary 2.3, $1, 2, \ldots, p-1$ are roots of $X^{p-1} - 1 \in \mathbb{Z}/p[X]$. Thus $(X-1)(X-2)\cdots$

$(X - (p-1)) \equiv X^{p-1} - 1 \pmod{p}$. Comparing constant terms, we deduce $(-1)^{p-1}(p-1)! \equiv -1 \pmod{p}$.

Equivalently, $(p-1)! \equiv (-1)^p \pmod{p}$. If p is odd, then $(-1)^p = -1$. If $p = 2$, then $(-1)^p \equiv -1 \pmod{p}$. Thus $(p-1)! \equiv -1 \pmod{p}$. ■

There is another approach to the main Lemma 1.4 which rests on the structure of the group U_p. Since it removes the mystery from Proposition 2.5 ($\Leftarrow$) we present it.

Theorem 2.7. U_p is a cyclic group for every prime p.

Definition. An integer x such that $\bar{x}$ generates U_p is called a *primitive root* mod p.

Proof. We use the following two facts.

Fact 1. For every d, the equation $X^d = 1$ has at most d distinct roots in U_p (by Lemma 2.6).

Fact 2. A cyclic group of order n contains a cyclic subgroup of every order d dividing n (generated by $y^{n/d}$ with y of order n).

Let $A(d)$ be the number of elements of order d in U_p. Since $|U_p| = p - 1$, we want to show that $A(p-1) > 0$, that is, that U_p contains at least one element of order $p - 1$ (which would then be a generator).

Let $C(d)$ be the number of elements of order d in a cyclic group of order $p - 1$.

Clearly $A(d) = C(d) = 0$ if d is not a divisor of $p - 1$. We note that

$$p - 1 = \sum_{\substack{\text{divisors} \\ d \text{ of } p-1}} A(d) = \sum_{\substack{\text{divisors} \\ d \text{ of } p-1}} C(d)$$

since the formulas on the right count the elements of U_p or of a cyclic group of order $p - 1$ by first grouping them by their orders.

Now suppose $A(d) \neq 0$. Then U_p contains an element of order d, which generates a cyclic subgroup of order d. All d elements of that subgroup solve $X^d = 1$, so by Fact 1 that subgroup must be the set of *all* roots of $X^d = 1$ in U_p; it must therefore contain all elements of order d in U_p. Therefore $A(d)$ equals the number of elements of order d in a cyclic group of order d. By Fact 2, this is less than or equal to $C(d)$.

We summarize. Either $A(d) = 0$ or $A(d) \le C(d)$. Either way, $A(d) \le C(d)$. But

$$\sum_{\substack{\text{divisors} \\ d \text{ of } p-1}} (C(d) - A(d)) = 0.$$

We conclude that $C(d) = A(d)$ for all d. In particular, $A(p-1) = C(p-1) > 0$. ■

Proof 2 of Lemma 1.4. Let p be an odd prime number. The equation $x^2 \equiv -1 \pmod{p}$ implies that $\bar{x} \in U_p$ has order 4, which implies that 4 divides $p-1$.

Conversely, suppose that 4 divides $p-1$. Let $z = y^{(p-1)/4}$ where y is a primitive root mod p. Then $\bar{z}$ has order 4 in U. Thus $z^2 \equiv -1 \pmod{p}$, because $\bar{z}^2$ is a root $\neq 1$ of $X^2 - 1$ in $\mathbb{Z}/p$. ■

Exercises

1. Show that $|U_{p^a}| = (p-1)p^{a-1}$ where p is prime, $a \ge 1$.

2. Let p be a prime number. Using the binomial formula for $(a+1)^p$, prove by induction that $a^p \equiv a \pmod{p}$ for every integer a.

3. Find the smallest prime divisor of $2^{37} - 1$.

4. Prove that every prime divisor of $2^{32} + 1$ is congruent to 1 mod 64. Find the five smallest primes congruent to 1 mod 64. Show that 641 divides $2^{32} + 1$.

5. Let $\mathrm{GCD}(a, m) = 1$. Prove that $a^d \equiv 1 \pmod{m}$, where $d = |U_m|$.

6. Let p be an odd prime number. Define sq: $U_p \to U_p$ by $sq(x) = x^2$. Show that sq is a group homomorphism. Calculate $|R|$ where R is the image of sq. By pairing x with x^{-1}, show that $R - \{1, -1\}$ contains an even number of elements. Conclude that $-1 \in R \Leftrightarrow p \equiv 1 \pmod{4}$. Hence prove Lemma 1.4.

7. How many roots of $X^2 - X$, of $X^3 - X$ are there in $\mathbb{Z}/6$? of $X^2 - 1$ in $\mathbb{Z}/8$?

8. Let p be prime. Show that the polynomial $X^m - 1$ has exactly m roots in $\mathbb{Z}/p$ for every positive divisor m of $p-1$.

9. Let G be a cyclic group of order n.

i. Show that every subgroup of G is cyclic and that G has exactly one subgroup of each order dividing n.

ii. Let $C(d)$ be the number of elements of G of order d. Show that $C(d) = |U_d|$ if $d|n$.

10. i. Let G be a finite group such that $X^d = 1$ has at most d solutions in G for every positive integer d. Show that G is cyclic.

ii. Let G be a finite subgroup of $K^\times$ where K is a field. Show that G is cyclic and that $G = \{x \in K | x^m = 1\}$ where $m = |G|$.

11. i. Let G be a finite abelian group. Let m be the largest order of an element of G. Prove that $x^m = 1$ for every $x \in G$.

ii. Let $G = U_p$, where p is prime, and let m be as in i. Use Lemma 2.6 to prove that $m = p - 1$. Hence prove Theorem 2.7.

12. Let p be a prime number and let y be a primitive root mod p. Prove that $(p-1)! \equiv y^{1+2+\cdots+(p-1)} \pmod{p}$. Conclude that $(p-1)! \equiv -1 \pmod{p}$.

13. Let p be a prime number that is congruent to 1 mod 8. Let $z = y^{(p-1)/8}$ where y is a primitive root mod p. Prove that $(z^3 - z)^2 \equiv 2 \pmod{p}$. Conclude that there exist integers a, b such that $a^2 \equiv 2 \pmod{p}$ and $b^2 \equiv -2 \pmod{p}$.

14. Let p be a prime congruent to 1 mod 3. Let $z = y^{(p-1)/3}$ where y is a primitive root mod p. Prove that $(2z+1)^2 \equiv -3 \pmod{p}$.

15. Let p be a prime congruent to 1 mod 5. Let $z = y^{(p-1)/5}$ where y is a primitive root mod p. Prove that $(2z^4 + 2z + 1)^2 \equiv 5 \pmod{p}$.

16. Prove that $\Sigma_{x=1}^{p-1} x^n \equiv 0 \pmod{p}$ for all prime numbers p and all integers $n \not\equiv 0 \pmod{p-1}$.

17. For what primes p does $X^3 \equiv a \pmod{p}$ have a solution for all integers a?

18. Let p be a prime number. Show that $(ap)!/p^a \equiv (-1)^a a! \pmod{p}$ for all positive integers a. Show that $p^2!/p^{p+1} \equiv 1 \pmod{p}$.

19. i. Let $a, n \in \mathbb{Z}$ with $n > 1$, $\text{GCD}(a, n) = \text{GCD}(n, 10) = 1$, and $1 \leq a < n$. Show that as a repeating decimal, $a/n = 0.\overline{x_1 x_2 \cdots x_r}$ with $x_i \in \{0, 1, \ldots, 9\}$ and minimal period r equal to the order of 10 in the group U_n.

ii. Find all primes p for which the decimal expansion of $1/p$ has period $r = 1, 2, 3, 4, 5$.

20. Let p be a prime number and let $n \geq 1 \in \mathbb{Z}$.

i. Let $a = 2^{p^{n-1}}$ and let q be a prime divisor of $(a^p - 1)/(a - 1) = a^{p-1} + a^{p-2} + \cdots + 1$. Show that $\bar{2} \in U_q$ has order equal to p^n and that $q \equiv 1 \pmod{p^n}$.

ii. Prove that there is an infinite number of primes congruent to 1 mod p^n.

iii. Prove that there is an infinite number of primes with final digit 1.

21. Let p be an odd prime number.

i. Let $t \in \mathbb{Z}$. Show that $(1 + tp)^{p^{n-2}} \equiv 1 + tp^{n-1} \pmod{p^n}$ for all $n \geq 2$. Conclude that if $p \nmid t$, then $1 + tp$ has order p^{n-1} in U_{p^n} for all $n \geq 1$.

ii. Show that U_{p^n} contains an element of order $p - 1$ for every $n \geq 1$.

iii. Prove that U_{p^n} is a cyclic group for all $n \geq 1$.

iv. An integer x is said to be a *primitive root* mod p^n iff $p \nmid x$ and $\bar{x}$ generates the group U_{p^n}. Show that 2 is a primitive root mod 3^n, 5^n, 11^n, and 13^n and that 3 is a primitive root mod 7^n for all $n \geq 1$.

v. Prove that if $x \in \mathbb{Z}$ is a primitive root mod p^2 then x is a primitive root mod p^n for all $n \geq 1$.

22. Prove that $5^{2^{n-3}} \equiv 1 + 2^{n-1} \pmod{2^n}$ for all $n \geq 3$. Deduce that $\bar{5}$ has order 2^{n-2} in U_{2^n} for all $n \geq 2$. Conclude that $U_{2^n} = \langle -\bar{1} \rangle \times \langle \bar{5} \rangle$ for $n \geq 3$. Thus if $n \geq 3$, then U_{2^n} is not cyclic.

23. Let G be a finite abelian group of order n. Use Proof 2 of Corollary 2.3 as a model to prove that $g^n = 1$ for all $g \in G$.

3. Applications of Lemma 1.4.

Lemma 3.1. Let n be an integer and let p be an odd prime dividing $n^2 + 1$. Then $p \equiv 1 \pmod 4$.

Proof. Since $p | n^2 + 1$, we have $n^2 \equiv -1 \pmod p$. By Lemma 1.4, $p \equiv 1 \pmod 4$. ■

Theorem 1.2. There are infinitely many primes congruent to 1 mod 4.

Proof. We imitate Euclid's proof of the infinitude of the primes. Let $p_1, p_2, \ldots, p_r$ be primes congruent to 1 mod 4. Let $n = 2p_1p_2 \cdots p_r$.

By Lemma 3.1, all prime divisors of $N = n^2 + 1$ (which is odd) must be congruent to 1 mod 4. Obviously, none of the p_i divide N. So the list $p_1, \ldots, p_r$ may be extended by tossing in *any* prime divisor of N. ■

We next prove the implication $1 \Rightarrow 2$ of Theorem 1.3.

Proposition 3.2. Let $x^2 + y^2 = n$ with integers x, y, n where $n \neq 0$. Let q be a prime congruent to 3 mod 4. Then the highest power of q dividing n is an even power.

Proof. Suppose that $q | n$. We first show that q must divide both x and y. If for instance $q \nmid x$, then x is invertible mod q. So the congruence $x^2 + y^2 \equiv 0 \pmod{q}$ would lead to $(yx^{-1})^2 \equiv -1 \pmod{q}$, which contradicts Lemma 1.4. Thus $q | x$. Similarly $q | y$.

Let q^a be the highest power of q dividing *both* x and y. Then $q^{2a} | n$ and we claim that $q^{2a+1} \nmid n$.

Write $X = x/q^a$, $Y = y/q^a$, and $N = n/q^{2a}$. We have $X^2 + Y^2 = N$. Since q does not divide both X and Y, the argument of the first half of this proof shows that q does not divide N. Thus the highest power of q dividing n is precisely q^{2a}. ■

Our next proposition, together with Theorem 1.1 (which has yet to be proved) and the trivial observations that $1^2 + 1^2 = 2$ and $p^2 + 0^2 = p^2$, serves to prove the implication $2 \Rightarrow 1$ of Theorem 1.3.

Proposition 3.3. If integers m, n are both sums of two (integer) squares, then so is their product mn.

Proof. Suppose $X^2 + Y^2 = m$ and $U^2 + V^2 = n$ with integers X, Y, U, V. We then have $m = |X + iY|^2$, $n = |U + iV|^2$. Thus

$$\begin{aligned} mn &= |(X + iY)(U + iV)|^2 = |(XU - YV) + i(XV + YU)|^2 \\ &= (XU - YV)^2 + (XV + YU)^2. \end{aligned}$$ ■

Exercises

1. Prove that $(X^2 + Y^2)(U^2 + V^2) = (XU + YV)^2 + (XV - YU)^2$. Find an interpretation of this identity in terms of absolute values of complex numbers.

2. Find four distinct positive integers a, b, c, d such that $a^2 + b^2 = c^2 + d^2 = 439097 = 577 \cdot 761$.

3. Let $m, n \in \mathbb{Z}$. Suppose that the equations $X^2 + 2Y^2 = m$, $X^2 + 2Y^2 = n$ both have integer solutions. Prove that the equation $X^2 + 2Y^2 = mn$ must also have a solution in integers.

4. Let $x^2 + 2y^2 = n$ with integers x, y, n where $n \neq 0$. Let q be a prime number such that there is no integer z satisfying the congruence $z^2 \equiv -2 \pmod{q}$. Prove that the highest power of q dividing n is an even power. Find the three smallest primes q to which this result applies.

5. **i.** Let p be an odd prime. Show that $X^4 \equiv -1 \pmod{p}$ has a solution if and only if $p \equiv 1 \pmod{8}$.

ii. Prove that there are infinitely many primes congruent to 1 mod 8.

6. Prove that there are infinitely many primes congruent to 5 mod 8. (*Hint:* Let P be a product of primes of the desired type. Consider prime divisors of $P^2 + 4$.)

7. Let $m, n \in \mathbb{Z}$ with $n \equiv -1 \pmod{4}$. Assume that m has no prime divisors that are congruent to 3 mod 4. Prove that there do not exist integers X, Y such that $Y^2 = X^3 + n^3 - 4m^2$. (Examples: $Y^2 = X^3 + k$, $k = -17, -5, 11, 23$.) (*Hint:* Show that if $Y^2 + 4m^2 = (X + n)(X^2 - nX + n^2)$, then there is a prime divisor of $X^2 - nX + n^2$ that is congruent to 3 mod 4. Produce a contradiction.)

8. Let m and n be odd integers. Assume that m has no prime divisors that are congruent to 3 mod 4. Prove that there do not exist integers X, Y such that $Y^2 = X^3 + 8n^3 - m^2$. (Examples: $Y^2 = X^3 + k$, $k = -9, 7$.)

4. Gaussian Integers

The factorization $X^2 + Y^2 = (X + iY)(X - iY)$ suggested to Gauss the study of the ring that now bears his name.

Definition. The *ring* $\mathbb{Z}[i]$ of *Gaussian integers* is the ring $\{m + in \mid m, n \in \mathbb{Z}\} \subset \mathbb{C}$.

The *norm* $N\alpha$ of a complex number α is defined by $N\alpha = |\alpha|^2$. For a Gaussian integer we have $N(m + in) = m^2 + n^2$.

Because $N\alpha = \alpha\bar{\alpha}$, we find that $N(\alpha\beta) = N\alpha \cdot N\beta$. From this we deduce that $\alpha \in \mathbb{Z}[i]$ is a unit in $\mathbb{Z}[i]$ if and only if $N\alpha = 1$. The group of units in $\mathbb{Z}[i]$ is thus $\{\pm 1, \pm i\}$.

Gauss discovered that much of Euclid's ancient theory of factorization of integers can be carried over to $\mathbb{Z}[i]$ with important number theoretic consequences. The cornerstone is Theorem 4.1, whose statement and proof is our next order of business.

Definition. A Gaussian integer $\alpha \neq 0$ is a *Gaussian prime* iff α is not a unit and there is no factorization $\alpha = \beta\gamma$ where $\beta, \gamma \in \mathbb{Z}[i]$ are nonunits.

Two Gaussian integers are *associate* iff one is a unit times the other.

Theorem 4.1. Let $\alpha \in \mathbb{Z}[i]$ with $\alpha \neq 0$. Then there exists a factorization $\alpha = up_1^{m_1} \cdots p_s^{m_s}$ where u is a unit in $\mathbb{Z}[i]$, the p_j are pairwise nonassociate Gaussian primes, and the m_j are positive integers. Moreover, if $\alpha = vq_1^{n_1} \cdots q_t^{n_t}$ is another such factorization, then $s = t$ and, after relabelling, p_j is associate to q_j and $m_j = n_j$ for all $j = 1, 2, \ldots, s$. (In short, $\mathbb{Z}[i]$ is a unique factorization domain.)

Proof of Theorem 4.1. Existence. We proceed by induction on $N\alpha$. If $N\alpha = 1$, we have already noted that α is a unit. If $N\alpha > 1$, then either α is a Gaussian prime or $\alpha = \beta\gamma$ with $N\beta, N\gamma > 1$. Since $N\alpha = N\beta N\gamma$, it follows that $N\beta, N\gamma < N\alpha$. By the inductive hypothesis, both β and γ have factorizations as products of Gaussian primes. These can be combined to produce the desired factorization of α. ∎

Before proving uniqueness of factorization we first prove that $\mathbb{Z}[i]$ possesses a division algorithm.

Proposition 4.2. Let $\alpha, \beta \in \mathbb{Z}[i]$ with $\beta \neq 0$. There exist $q, r \in \mathbb{Z}[i]$ such that $\alpha = q\beta + r$ and $Nr < N\beta$. (That is, we can divide α by β getting quotient q with a remainder r which is smaller than the divisor.)

Proof. Let $\alpha/\beta = \gamma \in \mathbb{C}$, say $\gamma = u + iv$ with $u, v \in \mathbb{R}$. Let m be the nearest integer to u, n the nearest integer to v, so that $\gamma = (m + s) + i(n + t)$ with $|s|, |t| \leq \frac{1}{2}$. Let $q = m + in \in \mathbb{Z}[i]$ and let $r = \beta(\gamma - q)$. Then $\alpha = q\beta + r$. Since $\alpha, q, \beta \in \mathbb{Z}[i]$, we have that $r = \alpha - q\beta \in \mathbb{Z}[i]$. Finally, $Nr = N\beta \cdot N(\gamma - q) = N\beta \cdot N(s + it) \leq N\beta(\frac{1}{2}^2 + \frac{1}{2}^2) < N\beta$. ∎

Definition. A nonzero Gaussian integer β is said to *divide* $\alpha \in \mathbb{Z}[i]$ (written $\beta|\alpha$) iff there exists $\gamma \in \mathbb{Z}[i]$ such that $\alpha = \beta\gamma$.

A Gaussian integer γ is a *greatest common divisor* (GCD) of a pair $\alpha, \beta \in \mathbb{Z}[i]$ iff γ divides both α and β and is itself divisible by all other elements of $\mathbb{Z}[i]$ with the same property.

Notice that if $\alpha|\beta$ and $\beta|\alpha$, then α and β are associate. For writing $\alpha = \beta\gamma$, we get $N\alpha = N\beta N\gamma$ and thus $N\beta \leq N\alpha$. Similarly, $N\alpha \leq N\beta$, whence $N\alpha = N\beta$ and so $N\gamma = 1$. Hence γ is a unit.

Lemma 4.3. i. Every pair of Gaussian integers not both zero has a greatest common divisor. Moreover, for each such pair $\alpha, \beta \in \mathbb{Z}[i]$ and greatest common divisor γ, there exist $x, y \in \mathbb{Z}[i]$ such that $\alpha x + \beta y = \gamma$.

ii. If a Gaussian prime divides a product $\alpha_1\alpha_2 \cdots \alpha_n$ of Gaussian integers, then it must divide one of the α_j.

Proof. i. By repeated use of Proposition 4.2 exactly as for the integers $\mathbb{Z}$, as follows. Let $\alpha, \beta \in \mathbb{Z}[i]$ with $\beta \neq 0$. If $\beta | \alpha$, then β is a GCD for α and β. If not, write inductively

$$\alpha = q_1\beta + r_1,$$
$$\beta = q_2 r_1 + r_2,$$
$$r_{m-1} = q_{m+1} r_m + r_{m+1},$$

where $q_m, r_m \in \mathbb{Z}[i]$ and $Nr_m < Nr_{m-1}$. Because the Nr_m are nonnegative integers, the process must stop with an integer M such that $Nr_{M+1} = 0$. From the definition of the norm we see that $r_{M+1} = 0$. It is then easy to prove that r_M is a GCD for α and β that can be expressed as a linear combination $\alpha x + \beta y$. If γ is another GCD for α and β, then $r_M | \gamma$ and $\gamma | r_M$, whence r_M and γ are associate. That is, $\gamma = u r_M$ for some unit $u \in \mathbb{Z}[i]$. We have $\alpha u x + \beta u y = \gamma$.

ii. Exercise. ∎

Proof of Theorem 4.1. Uniqueness. We just copy the classical proof of unique factorization in $\mathbb{Z}$ using Lemma 4.3ii. There are no surprises, so we leave the details as an exercise. ∎

The reader with a taste for abstraction might well prefer to replace the preceding argument with the proof of some such general theorem as Euclidean domains are unique factorization domains. He would then have to define Euclidean domains carefully, but his proof would be no different from that sketched.

We wish to point out the limited scope of the methods just sketched. Many rings are important in number theory; most do not have a unique factorization property. Of those that do, few possess a Euclidean algorithm. They must be tackled with completely new ideas, and in fact, there are proofs that factorization in the Gaussian integers is unique that in no way rely on Proposition 4.2. This is of very great interest but cannot be developed here.

We now produce our first proof of Theorem 1.1.

Proof of Theorem 1.1. Let p be a prime number congruent to 1 mod 4. By Lemma 1.4 there exists $x \in \mathbb{Z}$ such that $x^2 \equiv -1 \pmod{p}$. Then $p | x^2 + 1$. Since p divides a product $(x + i)(x - i)$ of two Gaussian integers but does not divide either factor $x + i$ or $x - i$, we conclude that p cannot be a Gaussian prime.

Thus p factors, $p = \alpha\beta$ where $\alpha, \beta \in \mathbb{Z}[i]$ and $N\alpha, N\beta > 1$ (neither α nor β a unit in $\mathbb{Z}[i]$). The calculation $p^2 = Np = N\alpha N\beta$ shows that the ordinary integers $N\alpha$ and $N\beta$ must both equal p. Write $\alpha = U + iV$. We have $p = N\alpha = U^2 + V^2$. ■

We proceed with the analysis of $\mathbb{Z}[i]$. Our next goal is the determination of the Gaussian primes.

Lemma 4.4. i. Every Gaussian prime divides some ordinary prime number.

ii. Let $\alpha \in \mathbb{Z}[i]$. If $N\alpha$ is a prime number, then α is a Gaussian prime.

Proof. i. Let α be a Gaussian prime. We have $\alpha\bar{\alpha} = N\alpha = \prod p_i$ where the p_i are ordinary prime numbers. Since α divides the product of the p_i, it must divide one of the p_i.

ii. Suppose $N\alpha = p$ is prime. Then $\alpha = \beta\gamma \Rightarrow N\alpha = N\beta N\gamma = p \Rightarrow$ either $N\beta = 1$ and β is a unit or $N\gamma = 1$ and γ is a unit. ■

To find all Gaussian primes we need only, according to Lemma 4.4i, factor all ordinary prime numbers into products of Gaussian primes.

Theorem 4.5. i. $2 = -i(1 + i)^2$ and $1 + i$ is a Gaussian prime.

ii. Let p be a prime number congruent to 1 mod 4 and let $p = X^2 + Y^2$ with $X, Y \in \mathbb{Z}$.

Then $p = (X + iY)(X - iY)$, and $X + iY$, $X - iY$ are nonassociate Gaussian primes.

iii. Every prime number that is congruent to 3 mod 4 is a Gaussian prime.

Thus there are two Gaussian primes corresponding to every prime number congruent to 1 mod 4, and one corresponding to each of the other prime numbers. (We say that $p \equiv 3 \pmod 4$ *remains prime*, that $p \equiv 1 \pmod 4$ *splits*, and that 2 *ramifies* in $\mathbb{Z}[i]$.)

Proof. i. by Lemma 4.4ii.

ii. That $X + iY$, $X - iY$ are Gaussian primes is a consequence of Lemma 4.4ii. The four associates of $X + iY$ are $X + iY$, $-X - iY$, $-Y + iX$, and $Y - iX$. None of these equals $X - iY$ since the formulas $X^2 + Y^2 = p$ shows that $X \neq 0$, $Y \neq 0$, and $X \neq \pm Y$.

iii. Suppose $p \equiv 3 \pmod 4$ and $p = \alpha\beta$ with nonunits α and β in $\mathbb{Z}[i]$. Then $p^2 = N\alpha N\beta$ implies that $N\alpha = p$. Writing $\alpha = X + iY$, we conclude that $p = X^2 + Y^2$ with $X, Y \in \mathbb{Z}$. Reducing mod 4 we get $3 \equiv X^2 + Y^2 \pmod 4$. But squares are all congruent to 0 or 1 mod 4. The sum of no two can be congruent to 3 mod 4, and so we have a contradiction. Thus p does not factor in $\mathbb{Z}[i]$. ■

Theorem 4.6. Let p be a prime number congruent to 1 mod 4 and let $p = X^2 + Y^2$ with $X, Y \in \mathbb{Z}$.

There are exactly eight distinct solutions $(U, V) \in \mathbb{Z}^2$ to the equation $U^2 + V^2 = p$, namely

$$(U, V) = (\pm X, \pm Y) \quad \text{and} \quad (U, V) = (\pm Y, \pm X).$$

Thus p is a sum of two squares in essentially only one way.

Proof. If $p = U^2 + V^2$, then $U + iV$ is a Gaussian prime factor of p. By Theorems 4.1 and 4.5ii it must be one of the four associates of $X + iY$ or one of the four associates of $X - iY$. These correspond to the eight distinct possibilities listed in the theorem. ■

Exercises

1. Write detailed proofs of Lemma 4.3ii and of uniqueness in Theorem 4.1.

2. Deduce Theorem 1.3 from Theorem 4.5.

3. Let n be a positive integer. Prove that there exist *relatively prime* integers a, b such that $a^2 + b^2 = n$ if and only if n is not a multiple 4 and n has no prime divisors congruent to 3 mod 4.

4. **i.** Let $\omega = -\frac{1}{2} + i(\sqrt{3}/2)$ and let $\mathbb{Z}[\omega] = \{m + n\omega | m, n \in \mathbb{Z}\} \subset \mathbb{C}$. Show that $\mathbb{Z}[\omega]$ is a ring whose only units are ± 1, $\pm\omega$, and $\pm\omega^2$.

   ii. Prove that for every $z \in \mathbb{C}$ there exists $q \in \mathbb{Z}[\omega]$ such that $N(z - q) < 1$. Prove for $\mathbb{Z}[\omega]$ the analogues of Proposition 4.2 and Lemma 4.3, and hence prove a unique factorization theorem for $\mathbb{Z}[\omega]$.

   iii. Prove that if p is an odd prime number for which $X^2 \equiv -3 \pmod{p}$ has a solution, then there exist integers x, y such that $x^2 + xy + y^2 = p$.

   iv. Let p be a prime number greater than 3. Prove that the following three assertions are equivalent. (a) $X^2 + XY + Y^2 = p$ has a solution in integers X, Y. (b) $X^2 \equiv -3 \pmod{p}$ has a solution. (c) $p \equiv 1 \pmod{3}$. (See Exercise 2.14.)

5. Prove that every element of $\mathbb{Z}[\omega]$ is associate to an element of the form $m + in\sqrt{3}$ with $m, n \in \mathbb{Z}$. For which prime numbers p do there exist $x, y \in \mathbb{Z}$ such that $x^2 + 3y^2 = p$?

6. **i.** Let α be a prime element of $\mathbb{Z}[\omega]$. Show that $N\alpha = p$ or $N\alpha = p^2$ for some prime number p in $\mathbb{Z}$. Deduce that exactly one of the following three statements is true: α is associate to $i\sqrt{3}$; α and $\bar{\alpha}$ are nonassociate; α is associate to a prime number in $\mathbb{Z}$.

ii. Suppose that $x, y \in \mathbb{Z}$, $\mathrm{GCD}(x, y) = 1$, $x \not\equiv y \pmod 2$, and $3 \nmid x$. Show that $\mathrm{GCD}'(x + iy\sqrt{3}, x - iy\sqrt{3}) = 1$ where GCD′ here means "greatest common divisor in $\mathbb{Z}[\omega]$."

iii. Let $x, y, z \in \mathbb{Z}$ be such that $x^2 + 3y^2 = z^3$ and $\mathrm{GCD}(x, y) = 1$. Prove that there exist integers a, b such that $x = a^3 - 9ab^2$ and $y = 3a^2b - 3b^3$. (*Hint:* Show that the hypotheses of ii are met. Using Exercise 5 conclude that $x + iy\sqrt{3} = u(a + ib\sqrt{3})^3$ for some $a, b \in \mathbb{Z}$ and unit $u \in \mathbb{Z}[\omega]$. Show that $u = \pm 1$.) Show that $\mathrm{GCD}(a, b) = 1$, $a \not\equiv b \pmod 2$, and $3 \nmid a$.

7. **i.** State and prove a unique factorization theorem for the ring $\mathbb{Z}[\sqrt{-2}] = \{m + in\sqrt{2} \mid m, n \in \mathbb{Z}\}$.

ii. Let y be an odd integer. Prove that $\mathrm{GCD}'(y + i\sqrt{2}, y - i\sqrt{2}) = 1$ where GCD′ here means "greatest common divisor in $\mathbb{Z}[\sqrt{-2}]$."

iii. Prove that $(x, y) = (3, \pm 5)$ are the only two integer solutions of the equation $y^2 = x^3 - 2$.

8. Let p and q be distinct primes congruent to 1 mod 4. Determine the number of pairs of integers (x, y) such that $0 < x < y$ and $x^2 + y^2 = pq$.

5. Farey Sequences

Consider the following array of fractions whose rows are the *Farey sequences* F_n.

$$
\begin{array}{llllllllllllll}
F_1 & \frac{0}{1} & & & & & & & & & & & & \frac{1}{1} \\[2ex]
F_2 & \frac{0}{1} & & & & & & \frac{1}{2} & & & & & & \frac{1}{1} \\[2ex]
F_3 & \frac{0}{1} & & & & \frac{1}{3} & & \frac{1}{2} & & \frac{2}{3} & & & & \frac{1}{1} \\[2ex]
F_4 & \frac{0}{1} & & & \frac{1}{4} & \frac{1}{3} & & \frac{1}{2} & & \frac{2}{3} & \frac{3}{4} & & & \frac{1}{1} \\[2ex]
F_5 & \frac{0}{1} & & \frac{1}{5} & \frac{1}{4} & \frac{1}{3} & \frac{2}{5} & \frac{1}{2} & \frac{3}{5} & \frac{2}{3} & \frac{3}{4} & \frac{4}{5} & & \frac{1}{1} \\[2ex]
F_6 & \frac{0}{1} & \frac{1}{6} & \frac{1}{5} & \frac{1}{4} & \frac{1}{3} & \frac{2}{5} & \frac{1}{2} & \frac{3}{5} & \frac{2}{3} & \frac{3}{4} & \frac{4}{5} & \frac{5}{6} & \frac{1}{1}
\end{array}
$$

The nth row F_n is the ascending sequence of integer fractions h/k between 0 and 1 with $\mathrm{GCD}(h, k) = 1$ and $1 \leq k \leq n$.

A little experimentation suggests that the difference of two successive elements of a Farey sequence is always a fraction with numerator 1. That this is actually so is the fundamental theorem of the subject.

Theorem 5.1. Let b/a, d/c be consecutive elements of the sequence F_n. Then $d/c - b/a = 1/ac$. Equivalently, $ad - bc = 1$.

The proof of Theorem 5.1 will be based on a geometric theorem which we discuss next.

Let $M \subset \mathbb{R}^2$ be the integer lattice of $\mathbb{R}^2$, that is, $M = \{(m, n) \in \mathbb{R}^2 | m, n \in \mathbb{Z}\}$. M is a group under vector addition.

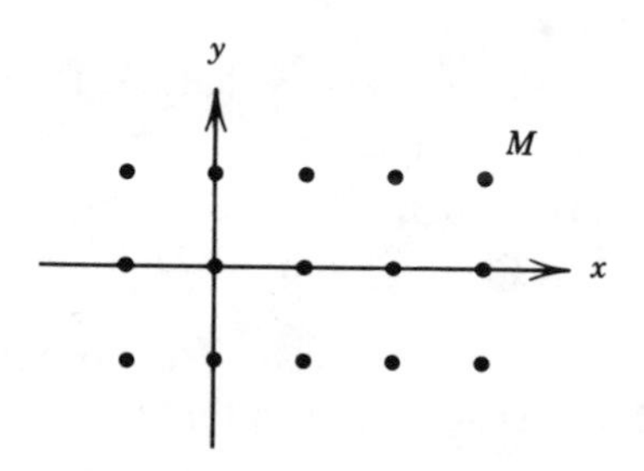

Definition. A point $P \in M$ is *visible* (from the origin O) iff $P \neq 0$ and there is no element of M on the interior of the segment OP. Equivalently,

$$P = (m, n) \in M \text{ is } \textit{visible} \text{ iff } \mathrm{GCD}(m, n) = 1.$$

Notice a bijection between the set of visible points $P = (x, y)$ of M in the region $0 \leq y \leq x$ and the set of rational numbers $q = y/x$ with $0 \leq q \leq 1$. The elements of the Farey sequence F_n correspond to the visible points (x, y) in the triangular region $0 \leq y \leq x \leq n$.

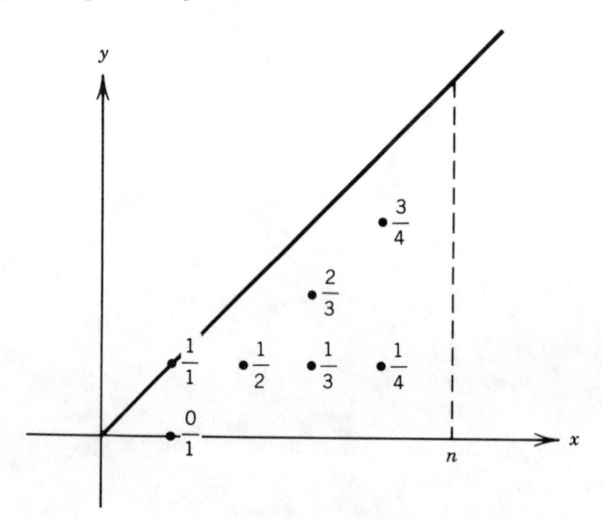

Definition. Let P, Q be two visible points of M with $P \neq \pm Q$. The *parallelogram* J defined by OP, OQ is the set $J = \{rP + sQ | r, s \in \mathbb{R}, 0 \leq r, s \leq 1\}$.

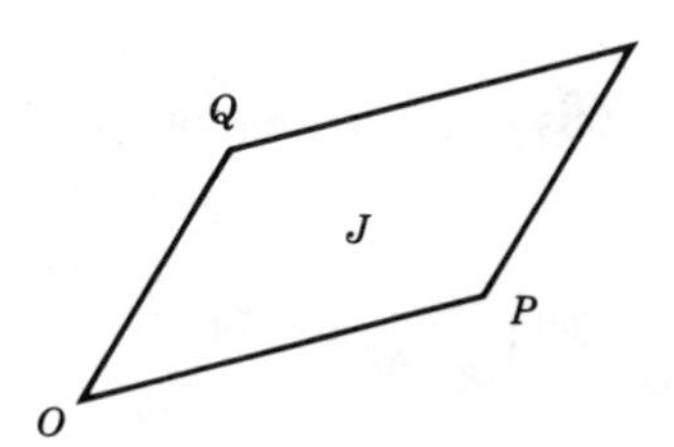

If $P = (a, b)$, $Q = (c, d)$, then from elementary geometry it is known that the *area* of J equals $|ad - bc|$, a positive integer.

The *triangle* OPQ is the set $OPQ = \{rP + sQ | r, s \in \mathbb{R};\ 0 \leq r, s \leq 1;\ r + s \leq 1\}$.

Theorem 5.2. Geometric Theorem. Let P, Q with $P \neq \pm Q$ be visible points of M and let δ be the area of the parallelogram J defined by OP, OQ.

If $\delta = 1$, then there is no point of M in the interior of J.

If $\delta > 1$, then there is at least one point of M in the intersection of OPQ with the interior of J.

We first prove a lemma.

Lemma 5.3. Let $P = (a, b)$, $Q = (c, d) \in M$. Let $M' = \{mP + nQ | m, n \in \mathbb{Z}\}$, the subgroup of M generated by P and Q.

The following are equivalent.

1. $M = M'$.
2. $(1, 0)$ and $(0, 1)$ are elements of M'.
3. $\det\begin{pmatrix} a & b \\ c & d \end{pmatrix} = \pm 1$.

Proof. Since $M' \subset M$, and M is generated by $\{(1,0), (0,1)\}$, the equivalence $1 \Leftrightarrow 2$ is obvious.

$2 \Rightarrow 3$. By hypothesis there exist integers m, n, p, q such that:

$$(1,0) = m(a, b) + n(c, d),$$

$$(0,1) = p(a, b) + q(c, d).$$

These equations can be written in matrix form:

$$\begin{pmatrix} 1 & 0 \\ 0 & 1 \end{pmatrix} = \begin{pmatrix} m & n \\ p & q \end{pmatrix} \begin{pmatrix} a & b \\ c & d \end{pmatrix}.$$

Taking the determinant of this last equation yields

$$1 = \det \begin{pmatrix} m & n \\ p & q \end{pmatrix} \cdot \det \begin{pmatrix} a & b \\ c & d \end{pmatrix}.$$

Since the two determinants are integers, each must equal ± 1.

$3 \Rightarrow 2$. The hypothesis $\det\begin{pmatrix} a & b \\ c & d \end{pmatrix} = \pm 1$ implies that the inverse of $\begin{pmatrix} a & b \\ c & d \end{pmatrix}$ has integer entries. Starting with

$$\begin{pmatrix} 1 & 0 \\ 0 & 1 \end{pmatrix} = \begin{pmatrix} m & n \\ p & q \end{pmatrix} \begin{pmatrix} a & b \\ c & d \end{pmatrix},$$

reverse the calculation in the proof of $2 \Rightarrow 3$ to show that $(1, 0), (0, 1) \in M'$. ■

Proof of Theorem 5.2. Let P, Q, J, and δ be as in the statement of the theorem.

Case 1. $\delta = 1$. Then Lemma 5.3 shows that $M = \{mP + nQ | m, n \in \mathbb{Z}\}$. But the interior of J is $\{rP + sQ | 0 < r, s < 1\}$. Clearly there is no intersection.

Case 2. $\delta > 1$. By Lemma 5.3 there is an element $R = xP + yQ$ of M that is not in M'. Because $R \notin M'$, x and y are not both integers. Let $S = R - [x]P - [y]Q$, where we write $[t]$ for the greatest integer less than or equal to a real number t.

Because S is a sum of three elements of the group M, we have $S \in M$. Clearly $S \in J$. If $S \in OPQ$, let $T = S$. If $S \notin OPQ$, let $T = P + Q - S$ which is an element of OPQ and also lies in $M \cap J$.

We show finally that T lies in the interior of J, that is, that $T \notin OP \cup OQ$. Since $T = rP + sQ$ with r, s not both integers we know that $T \neq O, P, Q$. That the element T of M is not in the interior of OP or OQ is because P and Q are visible. ■

Proof of Theorem 5.1. Let $b/a, d/c$ be consecutive fractions of F_n. Let $P = (a, b)$, $Q = (c, d)$.

If $\delta = |ad - bc| > 1$, then by Theorem 5.2 there is a visible point (k, h) of M other than P, Q inside OPQ. Slope considerations show that $b/a < h/k < d/c$. Since $k \leq \max\{a, c\} \leq n$, we find that $h/k \in F_n$. But that contradicts the consecutivity of b/a, d/c.

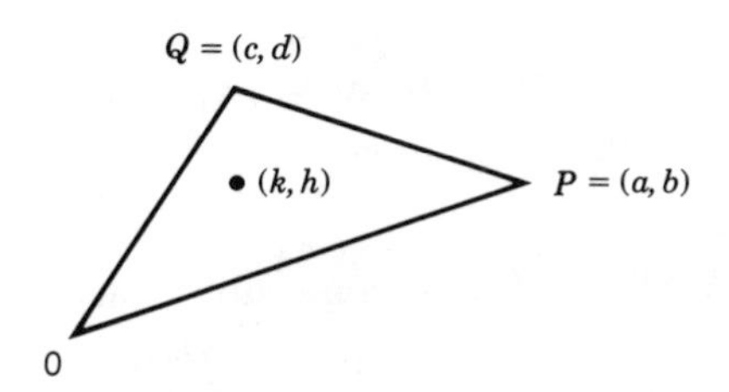

Thus $ad - bc = \pm 1$, where we have the positive sign because $b/a < d/c$. ■

We next employ Theorem 5.1 to analyze the construction of the Farey sequences.

Proposition 5.4. Let N_1/D_1, N_2/D_2 be consecutive fractions in F_n.

i. $\text{GCD}(N_1, N_2) = \text{GCD}(D_1, D_2) = 1$. That is, consecutive numerators and consecutive denominators from F_n are relatively prime.

ii. If $n \geq 2$, then $D_1 \neq D_2$. That is, consecutive denominators from F_n are unequal.

Proof. Part i is immediate from the equation $D_1 N_2 - N_1 D_2 = 1$.

In view of i, $D_1 = D_2$ can only happen if $D_1 = D_2 = 1$. But the only fractions in F_n with denominator 1 are $0/1$ and $1/1$, and for $n \geq 2$ they are not consecutive. ■

Corollary 5.5. Between two consecutive elements of F_n there is at most one element of F_{n+1}.

Proof. Fractions in F_{n+1} but not in F_n all have denominator $n + 1$. By Proposition 5.4ii any two such fractions must be separated by a fraction with a different and therefore smaller denominator. ■

Proposition 5.6. If N_1/D_1, N_2/D_2, N_3/D_3 are consecutive fractions in F_n, then $N_2/D_2 = (N_1 + N_3)/(D_1 + D_3)$.

Proof. By Theorem 5.1:

$$D_1N_2 - N_1D_2 = 1,$$

$$-D_3N_2 + N_3D_2 = 1.$$

Take the difference to get $(D_1 + D_3)N_2 - (N_1 + N_3)D_2 = 0$, which is equivalent to the assertion of the proposition. ■

Theorem 5.7. Let N_1/D_1, N_2/D_2 be consecutive fractions in F_n. Then

i. $D_1 + D_2 > n$ and $\text{GCD}(N_1 + N_2, D_1 + D_2) = 1$.

ii. $F_{D_1+D_2}$ is the first Farey sequence to include an element between N_1/D_1 and N_2/D_2. It includes exactly one such element, namely $(N_1 + N_2)/(D_1 + D_2)$.

Proof. An easy computation shows that $N_1/D_1 < (N_1 + N_2)/(D_1 + D_2) = A < N_2/D_2$.

Write A as a reduced fraction: $A = N/D$. Because the N_i/D_i are consecutive in F_n, $A \notin F_n$. Hence $D > n$. But D is a divisor of $D_1 + D_2$, and $D_1 + D_2 \leq 2n$, so $D = D_1 + D_2$. This proves i.

By Corollary 5.5 the first Farey sequence to include an element between N_1/D_1 and N_2/D_2 contains a unique such element. By Proposition 5.6 that element is $(N_1 + N_2)/(D_1 + D_2)$, which first occurs in $F_{D_1+D_2}$ by Theorem 5.7i. ■

The previous theorem gives a simple method to construct F_{n+1} from F_n. There is one new fraction $(N_1 + N_2)/(n + 1)$ with denominator $n + 1$ in F_{n+1} between every pair of consecutive fractions N_1/D_1, N_2/D_2 of F_n whose denominators add to $n + 1$.

Theorem 5.8. Let $N/D \in F_n$. The element of F_n that follows N/D is y/x where (x, y) is the unique pair of integers such that:

i.
$$Dy - Nx = 1$$

and

ii.
$$0 \leq n - D < x \leq n.$$

Proof. Let y_0/x_0 be the element of F_n that succeeds N/D. By Theorem 5.1, $Dy_0 - Nx_0 = 1$. By definition of F_n, $x_0 \le n$. By Theorem 5.7i, $x_0 + D > n$. Hence (x_0, y_0) is a solution to i and ii.

It remains but to observe that there is a *unique* solution to i and ii. Indeed, since $\text{GCD}(N, D) = 1$, the general solution in integers to $Dy - Nx = 1$ is the set $\{(x_0 + tD, y_0 + tN) | t \in \mathbb{Z}\}$. If $t > 0$, then $x_0 + tD > n$. If $t < 0$, then $x_0 + tD \le n - D$. So (x_0, y_0) is the only solution to i that satisfies ii as well. ∎

Farey sequences are fascinating in their own right. Our interest here, however, is in employing them to prove some simple results on the approximation of real numbers by rational numbers.

Let ξ be a real number. Let b and B be positive integers with $b < B$. Offhand, one might expect that it is possible to approximate ξ more closely by a fraction (rational number) with the large denominator B than by one with the small denominator b.

But that expectation is wrong! Counterexamples abound. For instance, the fraction $\frac{1}{9}$ is much closer to $\xi = 0.1111111$ than is $\frac{111}{1000}$, the best approximation possible with the large denominator 1000. More interesting perhaps is the approximation of π by $\frac{22}{7}$, which is closer to π than is 3.14, the nearest approximation by hundredths. A glance at the Farey sequence F_5 shows that every number between $\frac{1}{4}$ and $\frac{1}{3}$ will be closer to one of $\frac{1}{4}, \frac{1}{3}$ than to any approximation by fifths.

Just how closely can a real number ξ be approximated by a fraction with denominator b? It is easy to see that one can choose $a \in \mathbb{Z}$ such that $a/b - 1/2b \le \xi \le a/b + 1/2b$, thereby achieving $|\xi - a/b| \le 1/2b$. It turns out that for select b much better is possible, which is the basis of the examples of the preceding paragraph. The denominators 9, 7, and 3 or 4 there were very well adapted to the corresponding ξ.

Theorem 5.9. Let $\xi \in \mathbb{R}$ be irrational. There exist infinitely many rational numbers a/b such that $|\xi - a/b| < 1/2b^2$.

Proof. The calculation $|\xi - a/b| = |(\xi + k) - (a + bk)/b|$ shows that we may assume $0 < \xi < 1$.

Let m be a positive integer, and let n/d, N/D be the consecutive fractions in F_m such that $n/d < \xi < N/D$. We will prove that the inequality of the theorem is true for a/b one of n/d, N/D. This is trivial if $m = 1$, so we take $m \ge 2$. It readily follows from Theorem 5.1 and Proposition 5.4ii that

$$\left(\frac{n}{d} + \frac{1}{2d^2}\right) - \left(\frac{N}{D} - \frac{1}{2D^2}\right) = \frac{(D-d)^2}{2d^2D^2} > 0.$$

Therefore

$$\left(\frac{n}{d}, \frac{N}{D}\right) \subset \left(\frac{n}{d}, \frac{n}{d} + \frac{1}{2d^2}\right) \cup \left(\frac{N}{D} - \frac{1}{2D^2}, \frac{N}{D}\right).$$

Since $\xi \in (n/d, N/D)$, either $|\xi - n/d| < 1/2d^2$ or $|\xi - N/D| < 1/2D^2$.

We can produce infinitely many a/b with $|\xi - a/b| < 1/2b^2$ by using the procedure of the previous paragraph for different m. Indeed, we can extend any finite set S of a/b by choosing m so large that the consecutive fractions in F_m which bracket ξ are not in S. Such a choice is possible because the rational numbers are dense in the reals and each rational number between 0 and 1 is in F_m for large enough m. ■

Theorem 5.9 is not the best possible. One can in fact prove the existence of infinitely many fractions a/b such that $|\xi - a/b| < 1/\sqrt{5}\,b^2$.

It is occasionally useful to approximate a rational number by another rational with a smaller denominator. The following proposition will be strong enough for the application to Theorem 1.1.

Proposition 5.10. Let $\eta \geq 1$ be a real number and let $\xi \in [0, 1]$. There exists $a/b \in F_{[\eta]}$ such that $|\xi - a/b| < 1/b\eta$.

Proof. Let n/d, N/D be the consecutive fractions of $F_{[\eta]}$ such that $n/d < \xi < N/D$. Using Theorems 5.1 and 5.7i, calculate

$$\frac{n+N}{d+D} - \frac{n}{d} = \frac{1}{d(d+D)} \leq \frac{1}{d([\eta]+1)} < \frac{1}{d\eta},$$

$$\frac{N}{D} - \frac{n+N}{d+D} = \frac{1}{D(d+D)} \leq \frac{1}{D([\eta]+1)} < \frac{1}{D\eta}.$$

Clearly we may take a/b to be one of n/d, N/D. ■

We can now prove Theorem 1.1 a second time.

Proof of Theorem 1.1. Let p be a prime number congruent to 1 mod 4. By Lemma 1.4 there exists $x \in \mathbb{Z}$ such that $x^2 \equiv -1 \pmod{p}$. We may assume that $0 < x < p$.

Apply Proposition 5.10 to $\xi = x/p$ and $\eta = \sqrt{p}$. We find a fraction $a/b \in F_{[\sqrt{p}]}$ such that $|x/p - a/b| < 1/b\sqrt{p}$. Let $c = bx - ap = bp(x/p - a/b)$.

Note that $b^2 + c^2 \equiv b^2 + (bx)^2 \equiv 0 \pmod{p}$. In other words, $b^2 + c^2$ is a multiple of p.

Clearly $|c| < \sqrt{p}$. Since $a/b \in F_{[\sqrt{p}]}$, we also have $b < \sqrt{p}$. We conclude that $0 < b^2 + c^2 < 2p$. Necessarily, $b^2 + c^2 = p$. ∎

Exercises

1. Show that F_n contains $1 + \sum_{i=1}^{n} \phi(i)$ elements, where $\phi(i) = |U_i|$ for every $i \geq 1$.

2. Construct F_n for all $n \leq 11$.

3. Find the element of F_{30} that follows $\frac{5}{8}$.

4. An *Egyptian fraction* is a sum $\sum_{i=1}^{N}(1/m_i)$ where the m_i are distinct integers greater than 1. Show that every rational number q with $0 < q < 1$ equals an Egyptian fraction. Express $\frac{7}{19}$ as an Egyptian fraction.

5. Let ξ be an irrational number between 0 and 1.
   i. Let m, n/d, and N/D be as in the proof of Theorem 5.9. Show that $|\xi - a/b| < 1/\sqrt{5}\,b^2$ for a/b one of $n/d, N/D, (n+N)/(d+D)$. (*Hint:* If $D < ((\sqrt{5}-1)/2)d$ or $D > ((\sqrt{5}+1)/2)d$, then $n/d + 1/\sqrt{5}\,d^2 > N/D - 1/\sqrt{5}\,D^2$. If $((\sqrt{5}-1)/2)d < D < ((\sqrt{5}+1)/2)d$, then $n/d + 1/\sqrt{5}\,d^2 > (n+N)/(d+D) - 1/\sqrt{5}(d+D)^2$ and $(n+N)/(d+D) + 1/\sqrt{5}(d+D)^2 > N/D - 1/\sqrt{5}\,D^2$.)
   ii. Prove that there exist infinitely many rational numbers a/b such that $|\xi - a/b| < 1/\sqrt{5}\,b^2$.

6. Let ξ be a rational number. Prove that there are only finitely many pairs of integers a, b such that $0 < |\xi - a/b| < 1/b^2$.

7. Let m be a positive integer such that $X^2 \equiv -1 \pmod{m}$ has a solution. Prove that there are integers b and c such that $b^2 + c^2 = m$.

8. Let $\xi \in \mathbb{R}$ be irrational. Use Proposition 5.10 to prove that there exist infinitely many rational numbers a/b such that $|\xi - a/b| < 1/b^2$.

6. Minkowski's Theorem

Definition. The *lattice* $L = \langle v_1, v_2 \rangle$ determined by a basis $\{v_1, v_2\}$ of $\mathbb{R}^2$ is the subgroup of $\mathbb{R}^2$ generated by v_1 and v_2. That is, $L = \{mv_1 + nv_2 | m, n \in \mathbb{Z}\}$.

The *fundamental parallelogram* $D = D(L)$ of $L = \langle v_1, v_2 \rangle$ is the subset $D = \{xv_1 + yv_2 | x, y \in \mathbb{R};\ 0 \leq x,\ y < 1\}$ of $\mathbb{R}^2$.

The *area* $A = A(L)$ of L is the area of $D(L)$.

Notice that D is a set of representatives in $\mathbb{R}^2$ for the quotient group $\mathbb{R}^2/L$. That is, $\mathbb{R}^2 = \bigcup_{\gamma \in L}(\gamma + D)$ and the union is disjoint. (Exercise.)

A picture may help.

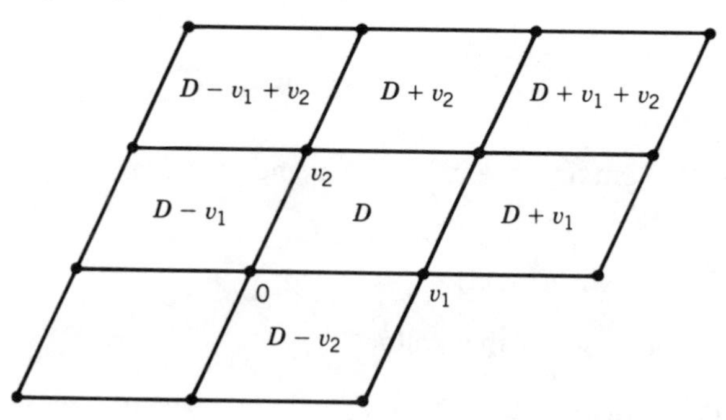

The basic tool in the geometry of numbers is:

Theorem 6.1. Minkowski's Theorem. Let $L = \langle v_1, v_2 \rangle$ be a lattice in $\mathbb{R}^2$ of area A. Let C be a circular disc centered at the origin of $\mathbb{R}^2$ and of area greater than $4A$. Then C contains a nonzero element of L.

Proof. The proof relies on a version of the box principle. If a region is cut into pieces and the pieces are laid into another region of lesser area, then there must be some overlapping of the pieces.

Define the subset $\tilde{D}$ of $\mathbb{R}^2$ by

$$\tilde{D} = D(2L) = \{xv_1 + yv_2 | x, y \in \mathbb{R}; 0 \leq x, y < 2\}.$$

The set $\tilde{D}$ has area $4A$ and is a set of representatives in $\mathbb{R}^2$ for the quotient group $\mathbb{R}^2/2L$. That is, $\mathbb{R}^2 = \bigcup_{\gamma \in L}(2\gamma + \tilde{D})$ and the union is disjoint.

Again, a picture may help.

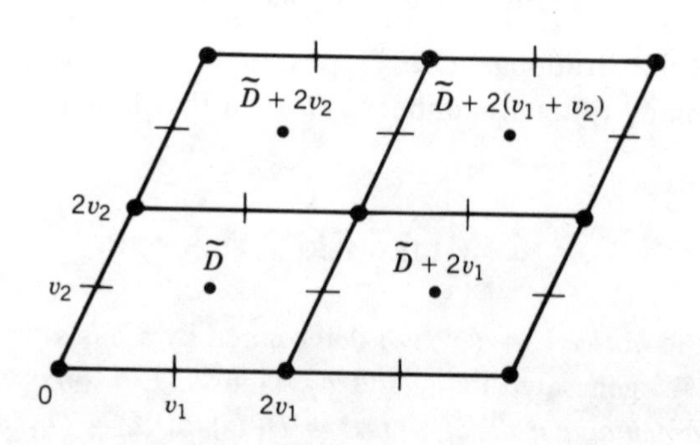

We now use $2L$ to transport, in pieces, the disc C to the parallelogram $\tilde{D}$. Precisely, for $\gamma \in L$ write $C_\gamma = (\tilde{D} + 2\gamma) \cap C$. We transport C_γ to $C_\gamma^{\text{trans}} = C_\gamma - 2\gamma \subset \tilde{D}$.

Since $C = \bigcup_{\text{disjoint}} C_\gamma$ and since the area of C_γ equals the area of C_γ^{trans}, we conclude that

$$\sum_{\gamma \in L} \left(\text{area of } C_\gamma^{\text{trans}}\right) = \text{area of } C > 4A = \text{area of } \tilde{D}.$$

Therefore, by the principle enunciated in the first paragraph of this proof, there must be some overlapping of the sets C_γ^{trans}; they cannot be pairwise disjoint. Thus there exists $\gamma \neq \gamma' \in L$, $P \in C_\gamma$, $Q \in C_{\gamma'}$ such that $P - 2\gamma = Q - 2\gamma'$.

Then $(P - Q)/2 = \gamma - \gamma' \neq 0 \in L$. Moreover, since $P, Q \in C$, then also $(P - Q)/2 \in C$ as the following picture shows. Thus $(P - Q)/2$ is the element of L sought.

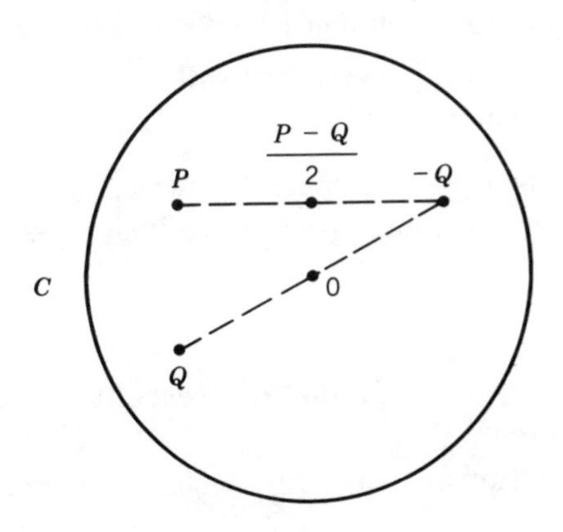

■

We give a third proof of Theorem 1.1.

Proof of Theorem 1.1. Let p be a prime number congruent to 1 mod 4. By Lemma 1.4 there exists $a \in \mathbb{Z}$ such that $a^2 \equiv -1 \pmod{p}$.

Form the lattice $L = \langle (p,0), (a,1) \rangle$, which has area $A = p$.

If $(x, y) \in L$, then $x^2 + y^2$ is an integer multiple of p. Indeed, $(x, y) = m(p,0) + n(a,1)$ for some $m, n \in \mathbb{Z}$ and $(mp + na)^2 + n^2 \equiv n^2a^2 + n^2 \equiv -n^2 + n^2 \equiv 0 \pmod{p}$.

Now let C be a disc centered at $(0,0)$ of radius less than $\sqrt{2p}$ and of area greater than $4A = 4p$. For example, we take C to be of radius $\frac{4}{3}\sqrt{p}$. By

Minkowski's Theorem there exists $(x, y) \neq 0 \in L \cap C$. Necessarily, $0 < x^2 + y^2 \leq r^2 = \frac{16}{9}p < 2p$. Since $x^2 + y^2$ is an integer multiple of p, we conclude that $x^2 + y^2 = p$. ∎

Exercises

1. **i.** Where in the proof of Theorem 6.1 did we use the hypothesis that C is a circular disc?

ii. A set $C \subset \mathbb{R}^2$ is *symmetric* iff $P \in C$ implies that $-P \in C$, and it is *convex* iff $P, Q \in C$ implies that the entire line segment PQ is contained in C. Prove that the assertion of Theorem 6.1 is true if the set C is a symmetric convex subset of $\mathbb{R}^2$ of area greater than $4A$.

2. Let $\eta, \xi \in \mathbb{R}$ with $\eta \geq 1$.

i. Prove that for every $\epsilon > 0$ there is a nonzero lattice point $(x, y) \in \mathbb{Z}^2$ such that $|y - \xi x| < 1/\eta$ and $|x| \leq \eta + \epsilon$. Deduce that there exists a nonzero $(x, y) \in \mathbb{Z}^2$ such that $|y - \xi x| < 1/\eta$ and $|x| \leq \eta$.

ii. Prove that there exist integers x, y with $1 \leq x \leq \eta$ such that $|\xi - y/x| < 1/x\eta$. (Alternative proof of Proposition 5.10.)

3. Prove that if m is a positive integer for which $X^2 \equiv -2 \pmod{m}$ has a solution, then there exist integers x, y such that $x^2 + 2y^2 = m$.

7. Method of Descent

Our third proof of Theorem 1.1 is possibly close to Fermat's own proof.

Proof of Theorem 1.1 by Descent. Let p be a prime number congruent to $1 \bmod 4$.

By Lemma 1.4 we choose $x \in \mathbb{Z}$ such that $x^2 \equiv -1 \pmod{p}$. We can and do assume that $0 \leq x \leq p - 1$. Thus $x^2 + 1^2 = mp$ for some $m \in \mathbb{Z}$ with $1 \leq m < p$.

So we can begin our descent with integers x, y, m such that

$$x^2 + y^2 = mp, \qquad 1 \leq m < p. \tag{7.1}$$

If $m = 1$, we are done. If $m > 1$, we will show how to modify x, y to X, Y such that $X^2 + Y^2 = m'p$ with $m' \in \mathbb{Z}$, $1 \leq m' < m$. If $m' = 1$, we will be done. If $m' > 1$, we repeat the procedure, modifying X, Y to get ever smaller positive multiples of p as sums of two squares. After finitely many steps, we must arrive at $X^2 + Y^2 = 1p$.

We describe the modification process.

Assume $m > 1$. Let $x' \equiv x,\ y' \equiv y \pmod{m}$ with $|x'|, |y'| \leq m/2$. We claim that

$$x'^2 + y'^2 = m'm \quad \text{with } m' \in \mathbb{Z} \text{ and } 1 \leq m' < m. \tag{7.2}$$

Indeed,

$$x'^2 + y'^2 \equiv x^2 + y^2 \equiv 0 \pmod{m},$$

$$x'^2 + y'^2 \leq \left(\frac{m}{2}\right)^2 + \left(\frac{m}{2}\right)^2 < m^2,$$

and $x'^2 + y'^2 \neq 0$, because the assumption that $x' = y' = 0$ implies $x \equiv y \equiv 0 \pmod{m}$, which implies (by (7.1)) that $m^2 | mp$, which is a contradiction since $1 < m < p$ and p is prime.

Now multiply (7.1) and (7.2) and reexpress to get

$$m^2m'p = (xx' + yy')^2 + (xy' - x'y)^2. \tag{7.3}$$

We calculate

$$xx' + yy' \equiv x^2 + y^2 \equiv 0 \pmod{m},$$

$$xy' - x'y \equiv xy - xy \equiv 0 \pmod{m}.$$

Therefore $X = (xx' + yy')/m$ and $Y = (xy' - x'y)/m$ are both integers. By (7.3) we have

$$X^2 + Y^2 = m'p \quad \text{as sought.}$$

■

Fermat himself developed, or perhaps invented, the method of descent. A sampling of the many interesting Diophantine equations to which he applied it is given in the exercises.

In operational terms, the method in its simplest form can be described as follows. One begins with a solution or hypothesized solution of a given Diophantine equation. A way is found to produce from it another solution that is in some sense smaller. Iterating, one finds smaller and smaller solutions.

In this manner, one is led either to a solution of a specified form or to a contradiction.

In the third proof of Theorem 1.1, the given equation was $X^2 + Y^2 = pM$, to be solved in integers. The size of a solution (X, Y, M) was measured by M and a solution was sought with $M = 1$.

When solution "size" is a positive integer M, the possibility of a contradiction arises. Namely, an *infinite* descent $M_1 > M_2 > M_3 > \cdots > 0$ is impossible. The Pythagorean proof that $\sqrt{2}$ is irrational, or equivalently that $X = Y = 0$ is the only integer solution of the equation $X^2 = 2Y^2$, produces a contradiction in just this way.

Sums of four squares can also be studied by descent. We prove next Lagrange's famous theorem.

Theorem 7.4. Lagrange's Four Squares Theorem. Every positive integer is a sum of four squares. In other words, for every positive integer n there exist integers x, y, z, w such that $x^2 + y^2 + z^2 + w^2 = n$.

We require two lemmas of Euler.

Lemma 7.5. Euler. If two integers are each sums of four squares, then so is their product.

Proof.

$$\begin{aligned}
&(x^2 + y^2 + z^2 + w^2)(x'^2 + y'^2 + z'^2 + w'^2) \\
&\quad = (xx' + yy' + zz' + ww')^2 + (xy' - yx' + wz' - zw')^2 \\
&\qquad + (xz' - zx' + yw' - wy')^2 + (xw' - wx' + zy' - yz')^2.
\end{aligned}$$ ∎

Lemma 7.6. Euler. For every odd prime number p there exist $x, y \in \mathbb{Z}$ such that $x^2 + y^2 \equiv -1 \pmod{p}$ and $0 \le x,\ y \le (p-1)/2$.

Proof. Let p be an odd prime number. The $(p+1)/2$ integers x^2 for $0 \le x \le (p-1)/2$ belong to $(p+1)/2$ distinct congruence classes mod p. Similarly, the integers $-1 - y^2$ for $0 \le y \le (p-1)/2$ belong to $(p+1)/2$ distinct congruence classes mod p. But the $p + 1$ numbers x^2 and $-1 - y^2$ cannot lie in distinct congruence classes mod p since there are only p such classes in all. For some x and some y we must have $x^2 \equiv -1 - y^2 \pmod{p}$. ∎

Proof of Theorem 7.4. From Lemma 7.5 we learn that it will be sufficient to prove that every prime number is a sum of four squares. This is because $1 = 1^2 + 0^2 + 0^2 + 0^2$ and every integer greater than 1 is a product of primes. After noting that $2 = 1^2 + 1^2 + 0^2 + 0^2$ we can restrict attention to odd primes.

Let p be an odd prime. We will prove that p is a sum of four squares.

With x and y as in Lemma 7.6 we have the equation $x^2 + y^2 + 1^2 + 0^2 = mp$ for some $m \in \mathbb{Z}$ with $1 \le m < p$. So we can begin our descent with integers x, y, z, w such that

$$x^2 + y^2 + z^2 + w^2 = mp, \qquad 1 \le m < p. \tag{7.7}$$

If $m = 1$, we are done. If $m > 1$, we will show how to modify x, y, z, w to X, Y, Z, W such that $X^2 + Y^2 + Z^2 + W^2 = m'p$ with $m' \in \mathbb{Z}$, $1 \le m' < m$. The proof is then completed just as was the proof of Theorem 1.1.

We describe the modification process. Assume $m > 1$.

If m is even, then an even number of the integers x, y, z, w are odd. After relabelling if necessary, then either all four are even, all four are odd, or x and y are even and z and w are odd. In all three cases we can write

$$\left(\frac{x+y}{2}\right)^2 + \left(\frac{x-y}{2}\right)^2 + \left(\frac{z+w}{2}\right)^2 + \left(\frac{z-w}{2}\right)^2 = \frac{m}{2} \cdot p.$$

So we take X, Y, Z, and W to be $(x \pm y)/2$ and $(z \pm w)/2$ with $m' = m/2$.

Now suppose that $m > 1$ is odd. Let $x' \equiv x$, $y' \equiv y$, $z' \equiv z$, and $w' \equiv w \pmod m$ with $|x'|, |y'|, |z'|, |w'| < m/2$. As in the proof of Theorem 1.1 we easily deduce that

$$x'^2 + y'^2 + z'^2 + w'^2 = m'm \quad \text{with } m' \in \mathbb{Z} \text{ and } 1 \le m' < m. \tag{7.8}$$

Multiplying (7.7) and (7.8) gives

$$\begin{aligned} m^2m'p &= \left(x^2 + y^2 + z^2 + w^2\right)\left(x'^2 + y'^2 + z'^2 + w'^2\right) \\ &= a^2 + b^2 + c^2 + d^2, \end{aligned} \tag{7.9}$$

where a, b, c, and d are given in the proof of Lemma 7.5. Upon remarking that a, b, c, and d are all divisible by m, we can take $X = a/m$, $Y = b/m$, $Z = c/m$, and $W = d/m$. It follows as desired that $X^2 + Y^2 + Z^2 + W^2 = m'p$ with $1 \le m' < m$. ■

Exercises

1. Let p be a prime number. Prove that $X = Y = Z = 0$ is the only integer solution of the equation $X^3 + pY^3 + p^2Z^3 = 0$.

2. **i.** Let p be a prime number such that $x^2 \equiv -2 \pmod p$ has a solution. Prove that there exist integers X, Y such that $X^2 + 2Y^2 = p$.

ii. Let p be a prime number congruent to 5 or 7 mod 8. Prove that there is no integer x such that $x^2 \equiv -2 \pmod{p}$.

iii. Let p be a prime number congruent to 7 mod 8. Prove that there is an integer x such that $x^2 \equiv 2 \pmod{p}$. (*Hint:* Express -1 and -2 as powers of a primitive root mod p.)

3. i. Let p be a prime number such that $x^2 \equiv 2 \pmod{p}$ has a solution. Prove that there exist integers X, Y such that $X^2 - 2Y^2 = p$. (*Hint:* After a descent closely modelled on that of the proof of Theorem 1.1, you will arrive at the equation $X^2 - 2Y^2 = m'p$ where $1 \leq |m'| < p$. If $m' < 0$, use the identity $(X + 2Y)^2 - 2(X + Y)^2 = -(X^2 - 2Y^2)$.)

ii. Let p be a prime number congruent to 3 or 5 mod 8. Prove that there is no integer x such that $x^2 \equiv 2 \pmod{p}$.

iii. Let p be an odd prime number. Prove that the following three assertions are equivalent. (a) $X^2 - 2Y^2 = p$ has a solution in integers X, Y. (b) $X^2 \equiv 2 \pmod{p}$ has a solution. (c) $p \equiv \pm 1 \pmod{8}$. (See Exercises 2.13 and 7.2iii.)

4. i. Let p be an odd prime number. Prove that the following three assertions are equivalent. (a) $X^2 + 2Y^2 = p$ has a solution in integers X, Y. (b) $X^2 \equiv -2 \pmod{p}$ has a solution. (c) $p \equiv 1$ or $3 \pmod{8}$. (See Exercises 2.13, 7.2, and 7.3ii.)

ii. Let n be a positive integer. Give a necessary and sufficient condition in terms of the prime factorization of n that there exist integers X, Y such that $X^2 + 2Y^2 = n$. (See Exercises 3.3 and 3.4.)

5. Let x, y, z be positive integers such that $x^2 + y^2 = z^2$, $\mathrm{GCD}(x, y) = 1$, and y is even. Show that $(z + x)/2$ and $(z - x)/2$ are square integers. Conclude that there exist positive relatively prime integers a, b of opposite parity such that $x = a^2 - b^2$, $y = 2ab$, and $z = a^2 + b^2$.

6. i. Prove that there do not exist three integers X, Y, Z with $XY \neq 0$ such that $X^4 + Y^4 = Z^2$. (*Sketch of solution:* Given positive integers X, Y, Z with $X^4 + Y^4 = Z^2$, construct positive integers x, y, z such that $x^4 + y^4 = z^2$ and $0 < z < Z$, then deduce a contradiction. If $\mathrm{GCD}(X, Y) > 1$, there is an easy construction, so assume that $\mathrm{GCD}(X, Y) = 1$ and that Y is even. Using Exercise 5, show that there exist positive integers a, b such that $X^2 = a^2 - b^2$, $Y^2 = 2ab$, and $Z = a^2 + b^2$. Similarly, find positive integers l, m such that $X = l^2 - m^2$, $b = 2lm$, and $a = l^2 + m^2$. From $(Y/2)^2 = alm$, conclude that a, l, m are squares, say $l = x^2$, $m = y^2$, $a = z^2$.)

ii. Prove that there do not exist integers X, Y, Z with $XY \neq 0$ such that $X^4 + Y^4 = Z^4$.

7. **i.** Prove that there do not exist integers X, Y, Z with $XY \neq 0$ such that $X^4 + Y^2 = Z^4$. (*Hint:* Take the same approach as in Exercise 6i. If $\text{GCD}(X, Y) = 1$ and Y is odd, the descent is quite similar. If $\text{GCD}(X, Y) = 1$ and Y is even, the descent is simpler, requiring only one application of Exercise 5.)

   ii. Prove that the equation $y^2 = x^3 - x$ has no rational solution $(x, y) \in \mathbb{Q}^2$ other than $y = 0$, $x = 0, \pm 1$. Sketch a graph of the curve $y^2 = x^3 - x$. (*Hint:* Let $r^2 = s^3 - s$ where $s = a/b$ is a reduced fraction with $b > 0$. Show that $ab(a^2 - b^2)$ is a square. Show that if $a > 0$, then there are integers x, y, z such that $a = x^2$, $b = y^2$, $a^2 - b^2 = z^2$. Thus $y^4 + z^2 = x^4$. Conclude that $s = 1$. Make a similar argument if $a < 0$.)

8. Prove that there do not exist three integers X, Y, Z with $XYZ \neq 0$ such that $X^3 + Y^3 = Z^3$. (*Sketch of solution:* Given X, Y, Z with $X^3 + Y^3 = Z^3$ and $XYZ \neq 0$, produce x, y, z with $x^3 + y^3 = z^3$ and $0 < |xyz| < |XYZ|$. This can be done as follows. Assume that $\text{GCD}(X, Y) = 1$ and that X and Y are odd. Write $X = p + q$, $Y = p - q$ with relatively prime integers p, q, and show that $2p(p^2 + 3q^2) = Z^3$. Case 1. If $3 \nmid p$, show that $2p$ and $p^2 + 3q^2$ are relatively prime and hence both cubes. Thus by Exercise 4.6iii there exist integers a, b with $p = a^3 - 9ab^2$, $q = 3a^2b - 3b^3$. From $2p = 2a(a - 3b)(a + 3b) =$ cube, conclude that there exist integers x, y, z such that $a - 3b = x^3$, $a + 3b = y^3$, $2a = z^3$. Case 2. If $3 | p$, then $6p \cdot (q^2 + 3(p/3)^2) = Z^3$ where the two factors are relatively prime. By Exercise 4.6iii, $p/3 = 3a^2b - 3b^3$. From $6p = 27 \cdot 2b(a - b)(a + b) =$ cube, show that there are integers x, y, z such that $2b = x^3$, $a - b = y^3$, $a + b = z^3$.)

9. Prove that the congruence $X^2 + Y^2 \equiv a \pmod{p}$ has a solution for every integer a and every prime p.

8. Reduction of Positive Definite Binary Quadratic Forms

Definition. *A real binary quadratic form* F (or just a *form*) is a polynomial in the two variables X, Y of the shape $F = F(X, Y) = aX^2 + bXY + cY^2$ with real coefficients a, b, c.

It is *integral* iff $a, b, c \in \mathbb{Z}$.

Its *discriminant* $\Delta = \Delta(F)$ is defined by the formula $\Delta = b^2 - 4ac$.

It is *positive definite* iff $\Delta < 0$, $a > 0$, and $c > 0$.

An integral form is said to *represent* an integer n iff the equation $aX^2 + bXY + cY^2 = n$ has a solution in integers X, Y.

The terminology positive definite is justified by the calculation

$$F(X,Y) = aX^2 + bXY + cY^2 = a\left(X + \frac{b}{2a}Y\right)^2 - \frac{\Delta}{4a}Y^2$$

which shows that precisely for positive definite forms, if $(x, y) \in \mathbb{R}^2$ and $(x, y) \neq (0,0)$, then $F(x, y) > 0$.

The polynomial $X^2 + Y^2$ is a positive definite integral binary quadratic form of discriminant -4. Fermat's Theorem 1.1 asserts that it represents every prime number that is congruent to 1 mod 4.

We wish to give a proof of Theorem 1.1 that exploits the fact that different integral forms of the same discriminant may represent exactly the same sets of integers. A simple example of this phenomenon is provided by the two forms $X^2 + Y^2$ and $(X + Y)^2 + Y^2 = X^2 + 2XY + 2Y^2$.

Definition. $\mathbf{GL}_2(\mathbb{Z})$ is the multiplicative group of 2×2 matrices $g = \begin{pmatrix} r & s \\ t & u \end{pmatrix}$ such that $r, s, t, u \in \mathbb{Z}$ and $\det g = \pm 1$.

$\mathbf{SL}_2(\mathbb{Z})$ is the subgroup of matrices in $\mathbf{GL}_2(\mathbb{Z})$ with determinant $+1$.

Definition. For a form $F = aX^2 + bXY + cY^2$ and a matrix $g = \begin{pmatrix} r & s \\ t & u \end{pmatrix} \in \mathbf{GL}_2(\mathbb{Z})$, we define the form gF by the formula

$$gF = a(rX + tY)^2 + b(rX + tY)(sX + uY) + c(sX + uY)^2.$$

That is, gF is gotten from F by making the substitution

$$X \mapsto rX + tY,$$
$$Y \mapsto sX + uY.$$

Lemma 8.1. i. The formula of the preceding definition is a group action of $\mathbf{GL}_2(\mathbb{Z})$ on the set of forms. That is, $\begin{pmatrix} 1 & 0 \\ 0 & 1 \end{pmatrix} F = F$ and $g_1(g_2F) = (g_1g_2)F$, where $g_1, g_2 \in \mathbf{GL}_2(\mathbb{Z})$ and F is a form.

ii. $\Delta(F) = \Delta(gF)$ for $g \in \mathbf{GL}_2(\mathbb{Z})$, F a form. That is, the action of $\mathbf{GL}_2(\mathbb{Z})$ on forms leaves the discriminant invariant.

iii. If F is positive definite or integral, then so is gF for all $g \in \mathbf{GL}_2(\mathbb{Z})$.

Proof. This lemma may be verified by straightforward calculations.

A more illuminating proof can be given by resorting to matrix notation as follows.

Write

$$F = aX^2 + bXY + cY^2 = (X \quad Y)\begin{pmatrix} a & \frac{b}{2} \\ \frac{b}{2} & c \end{pmatrix}\begin{pmatrix} X \\ Y \end{pmatrix}.$$

Note that

$$\Delta(F) = -4\det\begin{pmatrix} a & \frac{b}{2} \\ \frac{b}{2} & c \end{pmatrix}.$$

Let $g \in \mathbf{GL}_2(\mathbb{Z})$. Then

$$gF = (X \quad Y)g\begin{pmatrix} a & \frac{b}{2} \\ \frac{b}{2} & c \end{pmatrix}{}^t g\begin{pmatrix} X \\ Y \end{pmatrix},$$

where ${}^t g$ is the transpose of the matrix g.

Now i results from the equality ${}^t(g_1 g_2) = {}^t g_2 {}^t g_1$. Using the multiplicativity of the determinant, we calculate $\Delta(gF) = (\det g)\Delta(F)(\det {}^t g)$. Since $\det g = \det {}^t g$ and $(\det g)^2 = 1$, we find that $\Delta(gF) = \Delta(F)$, which is ii.

We leave iii as an exercise. ■

Definition. Two forms F and F' are *equivalent* iff there exists $g \in \mathbf{GL}_2(\mathbb{Z})$ such that $gF = F'$. If there exists $g \in \mathbf{SL}_2(\mathbb{Z})$ such that $gF = F'$, we say that F and F' are *properly equivalent*.

The equivalence (respectively, proper equivalence) classes of forms are just the orbits of the action of $\mathbf{GL}_2(\mathbb{Z})$ (respectively, $\mathbf{SL}_2(\mathbb{Z})$) on the set of forms. The relevance of form equivalence to number theory is given by Lemma 8.2.

Lemma 8.2. Integral forms that are equivalent represent precisely the same integers.

Proof. Let F be an integral form and let $g \in \mathbf{GL}_2(\mathbb{Z})$.

If $F(x, y) = n$, then $(gF)((x, y)g^{-1}) = n$, as is seen most directly in the matrix notation of the proof of Lemma 8.1. Clearly, if $(x, y) \in \mathbb{Z}^2$, then

$(x, y)g^{-1} \in \mathbb{Z}^2$ also. Thus gF represents every integer that F represents. A symmetric argument proves the reverse inclusion. ■

We begin the study of the set of proper equivalence classes of definite forms with the next proposition.

Proposition 8.3. i. Every positive definite integral form is properly equivalent to a form $F = aX^2 + bXY + cY^2$ such that $|b| \leq a \leq c$.

ii. If $|b| \leq a \leq c$ and $\Delta = b^2 - 4ac < 0$, then $a \leq \sqrt{|\Delta|/3}$.

Proof. A positive definite integral form $F = aX^2 + bXY + cY^2$ that does not satisfy $|b| \leq a \leq c$ can be modified within its proper equivalence class as follows. If $c < a$, "permute" X and Y by replacing F with $\begin{pmatrix} 0 & 1 \\ -1 & 0 \end{pmatrix} F = cX^2 - bXY + aY^2$. If $|b| > a$, replace F with $\begin{pmatrix} 1 & 0 \\ n & 1 \end{pmatrix} F = aX^2 + (b + 2an)XY + c'Y^2$ where $n \in \mathbb{Z}$ is chosen so that $|b + 2an| \leq a$. By alternating these two modification procedures we are led to a sequence of forms $F_n = a_nX^2 + b_nXY + c_nY^2$ such that $a_n \geq a_{n+1}$ and $a_n > a_{n+2}$. Since the a_n are all positive integers, the sequence must stop. This can only happen when $|b_n| \leq a_n \leq c_n$ as desired.

Part ii results from the calculation

$$4a^2 \leq 4ac = b^2 - \Delta \leq a^2 - \Delta.$$ ■

Corollary 8.4. There is only a finite number of proper equivalence classes of positive definite integral forms of a given discriminant.

Proof. By Proposition 8.3ii there are only finitely many positive definite integral forms $aX^2 + bXY + cY^2$ of discriminant Δ such that $|b| \leq a \leq c$. By Proposition 8.3i there is at least one in every proper equivalence class. ■

Corollary 8.5. All positive definite integral forms of discriminant -4 are properly equivalent.

Proof. We list the positive definite integral forms $aX^2 + bXY + cY^2$ of discriminant -4 such that $|b| \leq a \leq c$. By Proposition 8.3ii such a form satisfies $a \leq \sqrt{4/3}$. Hence $a = 1$ and $b = 0, \pm 1$. Since $\Delta = b^2 - 4c = -4$, b is even. Thus $b = 0$ and $c = 1$. So our list is the single form $X^2 + Y^2$.

Therefore, by Proposition 8.3i, all positive definite integral forms of discriminant -4 are properly equivalent to the same form $X^2 + Y^2$. Hence they form a single proper equivalence class. ■

We can now easily prove a strengthened version of Theorem 1.1.

Theorem 8.6. Every positive definite integral form of discriminant -4 (including $X^2 + Y^2$) represents every prime number that is congruent to 1 mod 4.

Proof. Let p be a prime number congruent to 1 mod 4. Since by Lemma 8.2 and Corollary 8.5 all positive definite integral forms of discriminant -4 represent the same integers, we need only prove the existence of a single such form that represents p.

By Lemma 1.4 there exists $m \in \mathbb{Z}$ such that $m^2 \equiv -1 \pmod{p}$. Then $(2m)^2 = -4 + np$ for some positive $n \in \mathbb{Z}$. Clearly np is a multiple of 4, from which it follows that n is a multiple of 4 as well. Thus the form $F(X, Y) = pX^2 + 2mXY + (n/4)Y^2$ is integral positive definite of discriminant -4. It represents p, since $F(1,0) = p$. So we have found what we needed. ■

We conclude Section 8 with a refinement of Proposition 8.3i that makes possible the enumeration of the proper equivalence classes of positive definite integral forms of a fixed discriminant.

Definition. A positive definite form $aX^2 + bXY + cY^2$ is *reduced* iff $|b| \leq a \leq c$; and in case $|b| = a$, then $b = a$; and in case $a = c$, then $b \geq 0$.

Theorem 8.7. There is a unique reduced form in every proper equivalence class of positive definite integral forms.

Proof. Existence. By Proposition 8.3i every positive definite integral form is properly equivalent to a form $F(X, Y) = aX^2 + bXY + cY^2$ such that $|b| \leq a \leq c$.

If $b = -a$, then $\begin{pmatrix} 1 & 0 \\ 1 & 1 \end{pmatrix} F$ is reduced. If $a = c$ and $b < 0$, then $\begin{pmatrix} 0 & 1 \\ -1 & 0 \end{pmatrix} F$ is reduced. Hence there is a reduced form in every proper equivalence class.

Uniqueness. Let $F(X, Y) = aX^2 + bXY + cY^2$ be a reduced positive definite integral form. We first prove that the following is a complete list of all solutions $(x, y) \in \mathbb{Z}^2$ of the inequality $0 < F(x, y) \leq a$:

8.8i. $\qquad (x, y) = (\pm 1, 0), \qquad F(x, y) = a,$

8.8ii. $\qquad a = c, \qquad (x, y) = (0, \pm 1), \qquad F(x, y) = a,$

8.8iii. $\qquad b = a = c, \qquad (x, y) = \pm(1, -1), \qquad F(x, y) = a.$

Indeed $ax^2 + bxy + cy^2 = a(x + (b/2a)y)^2 + (|\Delta|/4a)y^2 \le a$ implies $y^2 \le 4a^2/|\Delta| \le 4/3$, where the final inequality results from Proposition 8.3ii. Hence $y = 0, \pm 1$. If $y = 0$, then $x = \pm 1$ and $F(x, y) = a$ as in i. If $y = \pm 1$, we must solve $ax^2 \pm bx + c \le a$. Since $|b| \le a$ and $x \in \mathbb{Z}$, we have $ax^2 \pm bx \ge 0$. But $a \le c$, so a solution is possible only if $a = c$. Suppose that $a = c$ and let $y = \epsilon = \pm 1$. The equation to be solved is now $ax^2 + \epsilon bx = (ax + \epsilon b)x = 0$. This is easily analyzed, leading to ii and iii.

We conclude immediately that a is the smallest positive integer represented by the reduced form F.

Let now $F = aX^2 + bXY + cY^2$ and $F' = a'X^2 + b'XY + c'Y^2$ be two properly equivalent reduced positive definite integral forms. Necessarily $a = a'$, because a and a' can be described purely in terms of the integers represented by F and F', which are the same by Lemma 8.2.

Let $g = \begin{pmatrix} r & s \\ t & u \end{pmatrix} \in \mathbf{SL}_2(\mathbb{Z})$ satisfy $F' = gF$.

Suppose $c > a$. The calculation $a = a' = ar^2 + brs + cs^2$ together with (8.8) shows that $s = 0$. Since $\det g = 1$, we have $ru = 1$. Hence $b' = b + 2art \equiv b \pmod{2a}$. Since $|b|, |b'| \le a$, it follows that either $b = b'$ or $|b| = |b'| = a$. As F and F' are reduced, we get $b = b'$ either way. But F and F' have the same discriminant, so $a = a'$ and $b = b'$ imply that $c = c'$ as well. Thus $F = F'$.

Suppose that $c = a$. The preceding paragraph shows $c' > a'$ is impossible. Hence $a = a' = c = c'$. Discriminant considerations immediately prove that $b = \pm b'$. But since F and F' are reduced, $b \ge 0$ and $b' \ge 0$. Once again, we conclude that $F = F'$. ∎

Exercises

1. Let $\Delta \in \mathbb{Z}$. Prove that there is an integral form of discriminant Δ if and only if Δ is congruent to 0 or 1 mod 4.

2. Let $F(X, Y) = 10X^2 + 14XY + 5Y^2$. Find $g \in \mathbf{SL}_2(\mathbb{Z})$ such that $gF = X^2 + Y^2$. From g and the equation $4^2 + 9^2 = 97$, find all integer solutions (X, Y) to $F(X, Y) = 97$.

3. **i.** Prove that there is just one proper equivalence class of positive definite integral forms of discriminant -3. Show that if p is an odd prime for which $X^2 \equiv -3 \pmod p$ has a solution, then there exist integers x, y such that $x^2 + xy + y^2 = p$.

ii. Prove that if p is an odd prime for which $X^2 \equiv -7 \pmod p$ has a solution, then there exist integers x, y such that $x^2 + xy + 2y^2 = p$.

4. List all reduced positive definite integral forms of discriminants Δ such that $|\Delta| \le 23$.

5. Prove that in every equivalence class of positive definite integral forms there is a unique form $aX^2 + bXY + cY^2$ such that $0 \leq b \leq a \leq c$.

6. This exercise sketches an important geometric approach to Proposition 8.3 which applies to positive definite forms with real (not necessarily integral) coefficients. Let $H = \{\tau \in \mathbf{C} | \operatorname{Im} \tau > 0\}$ be the complex upper half plane.

i. Show that for positive definite $F(X, Y) = aX^2 + bXY + cY^2$ there is a unique $\tau = \tau(F) \in H$ such that $F(X, Y) = a(X - \tau Y)(X - \bar{\tau} Y)$. Prove that the map $F \to \tau$ is a bijection between the set of positive definite forms of a fixed discriminant and H.

ii. For $g = \begin{pmatrix} r & s \\ t & u \end{pmatrix} \in \mathbf{SL}_2(\mathbf{Z})$ and $\tau \in H$, let $g\tau = (r\tau + s)/(t\tau + u)$. Show that $g\tau \in H$. Prove that this definition of $g\tau$ gives a group action of $\mathbf{SL}_2(\mathbf{Z})$ on H.

iii. For F positive definite and $g \in \mathbf{SL}_2(\mathbf{Z})$, show that $\tau(gF) = {}^t g^{-1} \tau(F)$. Prove that two positive definite forms F and F' of the same discriminant are properly equivalent if and only if $\tau(F)$ and $\tau(F')$ are in the same $\mathbf{SL}_2(\mathbf{Z})$ orbit.

iv. Let $\tau \in H$. Show that there is an element τ' of the $\mathbf{SL}_2(\mathbf{Z})$ orbit of τ that has maximal imaginary part. Show that there exists $n \in \mathbf{Z}$ such that $\tau'' = \begin{pmatrix} 1 & n \\ 0 & 1 \end{pmatrix} \tau'$ satisfies $|\operatorname{Re} \tau''| \leq \frac{1}{2}$. By considering $\begin{pmatrix} 0 & -1 \\ 1 & 0 \end{pmatrix} \tau''$, prove that $|\tau''| \geq 1$. Conclude that every $\mathbf{SL}_2(\mathbf{Z})$ orbit in H contains at least one element $\tau \in D = \{z \in H \,|\, |z| \geq 1, |\operatorname{Re} z| \leq \frac{1}{2}\}$. Sketch D.

v. Prove that every proper equivalence class of positive definite real binary quadratic forms contains at least one form $aX^2 + bXY + cY^2$ such that $|b| \leq a \leq c$.

7. Let Δ be a positive integer that is not a square.

i. By imitating the proof of Proposition 8.3, show that every integral form of discriminant Δ is properly equivalent to a form $aX^2 + bXY + cY^2$ such that $|b| \leq |a| \leq |c|$, and that such a form satisfies $ac < 0$, $|a| \leq \frac{1}{2}\sqrt{\Delta}$.

ii. Prove that there is only a finite number of proper equivalence classes of integral forms of discriminant Δ.

iii. Show that every integral form of discriminant $\Delta = 5$ (resp. $\Delta = 8$) is properly equivalent to $X^2 + XY - Y^2$ (resp. $X^2 - 2Y^2$).

iv. Prove that if p is an odd prime for which $X^2 \equiv 5 \pmod{p}$ has a solution, then there exist integers x, y such that $x^2 + xy - y^2 = p$.

8. Let m be a positive integer, and let $\Delta = m^2$.

i. Show that for every integral form $F(X, Y)$ of discriminant Δ there exist integers x, y not both zero such that $F(x, y) = 0$.

ii. Show that every integral form of discriminant Δ is properly equivalent to a form $F = aX^2 + bXY$ where $|b| = m$. By consideration of gF where $g = \begin{pmatrix} r & s \\ t & u \end{pmatrix} \in \mathbf{SL}_2(\mathbb{Z})$, $t = m/d$, $u = a/d$, and $d = \mathrm{GCD}(m, a)$, show that one can take $b = +m$ in the preceding sentence.

iii. Show that every proper equivalence class of integral forms of discriminant Δ contains a unique form $aX^2 + mXY$ such that $-m/2 < a \leq m/2$. Thus there are exactly m such proper equivalence classes.

9. Prove that every integral form of discriminant $\Delta = 0$ is properly equivalent to a unique form $F = aX^2$.

10. Prove that the two forms $X^2 + 3Y^2$ and $X^2 + XY + Y^2$ represent the same integers, but that they are not equivalent.

11. Write three computer programs to perform as follows.

i. Input: A positive definite integral form f.
Task: To produce $g \in \mathbf{SL}_2(\mathbb{Z})$ such that gf is reduced.

ii. Input: Two positive definite integral forms f and f'.
Task: To determine whether f and f' are properly equivalent and if so to produce $g \in \mathbf{SL}_2(\mathbb{Z})$ such that $gf = f'$.

iii. Input: A negative integer Δ.
Task: To list all reduced positive definite integral forms of discriminant Δ.

12. Prove that every integral form of discriminant $\Delta < 0$ represents some integer $n \neq 0$ such that $|n| \leq \sqrt{|\Delta|/3}$.

13. Prove that every integral form of nonsquare discriminant $\Delta > 0$ represents some integer $n \neq 0$ such that $|n| \leq \frac{1}{2}\sqrt{\Delta}$. (See Exercise 7.)

14. Prove that for every $g \in \mathbf{GL}_2(\mathbb{Z})$ of order 4 (resp. order 3) there exists $\gamma \in \mathbf{GL}_2(\mathbb{Z})$ such that $\gamma g \gamma^{-1} = \begin{pmatrix} 0 & -1 \\ 1 & 0 \end{pmatrix}$ $\left(\text{resp. } \gamma g \gamma^{-1} = \begin{pmatrix} -1 & -1 \\ 1 & 0 \end{pmatrix}\right)$.

CHAPTER 3
Quadratic Reciprocity

1. Introduction

In this chapter we solve the problem: Given an integer Δ, describe the set of prime numbers that can be represented by an integral binary quadratic form of discriminant Δ.

We solved this problem for discriminant $\Delta = -4$ during our study of the form $X^2 + Y^2$. The set of primes p in question in that case is describable by congruence conditions, namely, $p = 2$ and $p \equiv 1 \pmod 4$. A similar thing is true in the general case though it is much harder to prove.

The first step is easy.

Proposition 1.1. Let $\Delta \in \mathbb{Z}$ be congruent to 0 or 1 mod 4, and let p be an odd prime number. Then the following two assertions are equivalent.

1. There is an integral binary quadratic form of discriminant Δ that represents p.
2. The congruence $X^2 \equiv \Delta \pmod p$ has a solution.

A thorough study of quadratic congruences is clearly suggested. The successful completion of such a study by the teenage Gauss is one of the great stories of mathematics. It will be the subject of the present chapter.

The key technical lemma, which was discovered by Euler and proved by Gauss, is known as the Law of Quadratic Reciprocity.

Proposition 1.2. The Law of Quadratic Reciprocity. Let p and q be distinct odd primes.

Then $X^2 \equiv q \pmod{p}$ has a solution if and only if $X^2 \equiv p \pmod{q}$ has a solution, unless $p \equiv q \equiv 3 \pmod{4}$, in which case one of the two congruences has a solution and one does not.

It is just a short hop to the definitive statement.

Theorem 1.3. Let a be an integer that is not a square.

There exists a surjective group homomorphism $\chi\colon U_{4a} \to \{\pm 1\}$ such that the following two conditions are equivalent for all odd primes p that do not divide a.

1. $X^2 \equiv a \pmod{p}$ has a solution.
2. $\chi(\bar{p}) = 1$.

The remarkable feature of Theorem 1.3 is that the condition that $X^2 \equiv a \pmod{p}$ be solvable can be stated as a congruence condition $(\bmod\, 4a)$ on the prime p. We have already seen this in the case $a = -1$. The existence of the homomorphism χ, called the Kronecker symbol, is so important that statements like Theorem 1.3 and its generalizations have themselves come to be called reciprocity laws. Classical quadratic reciprocity Proposition 1.2 can be easily deduced from Theorem 1.3.

In Section 10 we will prove Proposition 1.1 and combine it with Theorem 1.3 to give a complete solution to the motivating problem of the representability of primes by binary forms of a given discriminant.

There is a discussion of quadratic congruences with composite moduli, including the widely useful Chinese Remainder Theorem, in Section 2.

In Section 3 we introduce the Legendre symbol, which makes it possible to formulate the Law of Quadratic Reciprocity as an equation.

Gauss discovered eight proofs of the Law of Quadratic Reciprocity. In this chapter we present variants of his first (1796), third (1808), fourth (1805), and sixth (1818) proofs. (His second (1801) proof will be given in Chapter 5.) I like them all.

In Section 4 we give Gauss's and the world's first proof. It has been called ugly, but I do not agree. It is a computationally based induction on the maximum of the two primes p and q, which strikes me as quite natural.

Textbook tradition has declared that the simplest proof of quadratic reciprocity is the variant of Gauss's third proof that we reproduce in Section 5. It is a combinatorial proof based on one of the many famous Gauss Lemmas.

Routes to quadratic reciprocity via the trigonometric sums now called Gauss sums have proved more suggestive than either of the preceding. We explore two such in Sections 6 and 7. The proof in Section 7 is perhaps the most significant, pointing as it does toward contemporary algebraic number theory from higher reciprocity to class field theory. It is also very simple.

Section 8 emphasizes the practical aspects of the reciprocity law. We discuss the evaluation of the Legendre symbol. An interesting digression on a probabilistic method to search for large prime numbers is included.

The Kronecker symbol is constructed in Section 9. With it comes the proof of Theorem 1.3.

Exercises

1. Let $p \neq 5$ be a prime number. Deduce from Proposition 1.2 that the congruence $X^2 \equiv 5 \pmod{p}$ had a solution if and only if $p \equiv 1$ or $4 \pmod{5}$.

2. Determine explicitly the homomorphism χ of Theorem 1.3 for the case $a = 3$.

2. Composite Moduli

The theory of congruences is the study of equations in the rings $\mathbb{Z}/m$. It is in some respects simplest for prime moduli m, for in that case $\mathbb{Z}/m$ is a field and techniques from geometry can be applied most easily. Much is known, but more is conjectured.

Two important theorems, the Chinese Remainder Theorem and Hensel's Lemma (Newton's Method), sometimes together reduce questions in composite moduli to simpler questions in prime moduli. We will prove the first of these theorems and present a special case of the second in this section.

We begin with the possibility of reducing to prime power modulus.

Proposition 2.1. Let $m_1, m_2, \ldots, m_r \in \mathbb{Z}$ be positive and pairwise relatively prime.

i. The following two conditions are equivalent for every pair of integers x and y.

1. $x \equiv y \pmod{m_1 m_2 \cdots m_r}$.
2. $x \equiv y \pmod{m_i}$ for all $i = 1, 2, \ldots, r$.

ii. $\bigcap_{i=1}^{r} m_i\mathbb{Z} = m_1 m_2 \cdots m_r\mathbb{Z}$.

Proof. It is clearly enough to prove i in the case $y = 0$, which is identical with ii.

Suppose $x \equiv 0 \pmod{m_i}$ for all i. We will prove that $x \equiv 0 \pmod{m_1 m_2 \cdots m_r}$ by induction on r. The first case is $r = 2$. Since $\mathrm{GCD}(m_1, m_2) = 1$, there are $a, b \in \mathbb{Z}$ such that $am_1 + bm_2 = 1$. We have $am_1x + bm_2x = x$, where the terms on the left are both divisible by m_1m_2, the first because $m_2|x$ and the second because $m_1|x$. Thus $m_1m_2|x$. If $r > 2$, apply an inductive hypothesis to the sequence $m_1m_2, m_3, \ldots, m_r$ of $r - 1$ pairwise relatively prime divisors of x. ■

The comparison of the rings $\mathbb{Z}/m$ with different m inevitably involves consideration of homomorphisms between them. Let d and m be nonzero integers such that $d|m$. There is a containment $d\mathbb{Z} \supset m\mathbb{Z}$ of ideals of $\mathbb{Z}$. In the usual way this gives a surjective ring homomorphism $\mathbb{Z}/m \to \mathbb{Z}/d$ that maps $\bar{x} = x + m\mathbb{Z} \in \mathbb{Z}/m$ to $\bar{x} = x + d\mathbb{Z} \in \mathbb{Z}/d$ for every $x \in \mathbb{Z}$. This homomorphism is the subject of the Chinese Remainder Theorem.

Theorem 2.2. Chinese Remainder Theorem. Let $m_1, m_2, \ldots, m_r \in \mathbb{Z}$ be positive and pairwise relatively prime.

i. Let $a_1, a_2, \ldots, a_r \in \mathbb{Z}$. There exists $x \in \mathbb{Z}$ such that $x \equiv a_i \pmod{m_i}$ for $1 \leq i \leq r$.

ii. The map $f\colon \mathbb{Z}/m_1m_2 \cdots m_r \to \mathbb{Z}/m_1 \times \mathbb{Z}/m_2 \times \cdots \times \mathbb{Z}/m_r$, $f(\bar{x}) = (\bar{x}, \bar{x}, \ldots, \bar{x})$, is an isomorphism of rings.

Proof. i. Let $n_i = (m_1m_2 \cdots m_r)/m_i$. Then $\mathrm{GCD}(n_i, m_i) = 1$, so there exists $y_i \in \mathbb{Z}$ such that $n_iy_i \equiv 1 \pmod{m_i}$. Let $x_i = n_iy_i$. We have the congruences $x_i \equiv 1 \pmod{m_i}$ and $x_i \equiv 0 \pmod{m_j}$ for $j \neq i$. Therefore, $x = a_1x_1 + a_2x_2 + \cdots + a_rx_r \equiv a_i \pmod{m_i}$ for all i.

ii. Surjectivity by Theorem 2.2i and injectivity by Proposition 2.1. Observe that we actually constructed the inverse of f in the proof of i. ■

It is worth noting that the existence of a bijection between the rings of Theorem 2.2ii that is required to preserve only the *additive* structures is even simpler to prove. Such additive isomorphisms are not unique. The following will be useful to us later.

Proposition 2.3. Let m, n be nonzero relatively prime integers.

The function $f\colon \mathbb{Z}/m \oplus \mathbb{Z}/n \to \mathbb{Z}/mn$, $f(\bar{x}, \bar{y}) = \overline{nx + my}$, is an isomorphism of additive abelian groups.

Proof. Note that f is indeed well defined. It is also easily verified that f is a homomorphism.

Let $z = \bar{a} \in \mathbb{Z}/mn$. By Proposition 4.1 of Chapter 1 there exist $x, y \in \mathbb{Z}$ such that $nx + my = a$. Therefore, $z = \bar{a}$ lies in the image of f. We have proved that f is surjective.

Finally, suppose that $f(\bar{x}, \bar{y}) = 0$. This means that $nx + my \in mn\mathbb{Z}$. Then Lemma 3.6 of Chapter 1 implies that $x \in m\mathbb{Z}$ and that $y \in n\mathbb{Z}$. Therefore f is injective. ■

The next proposition is the application of the Chinese Remainder Theorem to the reduction of modulus in congruence equations.

Proposition 2.4. Let $f(X) \in \mathbb{Z}[X]$. Let $m_1, \dots, m_r$ be positive pairwise relatively prime integers. Then the following two conditions are equivalent.

1. $f(X) \equiv 0 \pmod{m_1 m_2 \cdots m_r}$ has a solution.
2. $f(X) \equiv 0 \pmod{m_i}$ has a solution for $i = 1, 2, \dots, r$.

Proof. $1 \Rightarrow 2$. Every solution $x \in \mathbb{Z}$ of congruence 1 is a solution of all the congruences 2.

$2 \Rightarrow 1$. Suppose that all the congruences of 2 have solutions. For each i let $x_i \in \mathbb{Z}$ be such that $f(x_i) \equiv 0 \pmod{m_i}$. By Theorem 2.2i choose $x \in \mathbb{Z}$ such that $x \equiv x_i \pmod{m_i}$ for $1 \leq i \leq r$. Then $f(x) \equiv 0 \pmod{m_i}$ for all i, and thus by Proposition 2.1i we have $f(x) \equiv 0 \pmod{m_1 m_2 \cdots m_r}$. ■

In most applications of Proposition 2.4 the integers m_i will be powers of distinct primes. It is natural to wonder whether a further reduction from prime power modulus to prime modulus is possible. Rather than address this question in full generality, we choose to treat the special case of quadratic congruences which is all that we will need in the sequel.

Proposition 2.5. i. Let $a \in \mathbb{Z}$ and let p be an odd prime not dividing a. Then the congruence $X^2 \equiv a \pmod{p^n}$ has a solution for every $n \geq 1$ if and only if $X^2 \equiv a \pmod{p}$ has a solution.

Moreover, if solutions exist, then $X^2 = \bar{a} \in \mathbb{Z}/p^n$ has exactly two solutions in $\mathbb{Z}/p^n$ for every $n \geq 1$.

ii. Let a be an odd integer. Then the congruence $X^2 \equiv a \pmod{2^n}$ has a solution for every $n \geq 3$ if and only if $X^2 \equiv a \pmod{8}$ has a solution if and only if $a \equiv 1 \pmod{8}$.

Moreover, if solutions exist, then $X^2 = \bar{a} \in \mathbb{Z}/2^n$ has exactly four solutions in $\mathbb{Z}/2^n$ for every $n \geq 3$.

Proof. i. Let $n \geq 1$ and let $x \in \mathbb{Z}$ satisfy $x^2 \equiv a \pmod{p^n}$. We prove that there exists $y \in \mathbb{Z}$ such that $y^2 \equiv a \pmod{p^{n+1}}$ and $y \equiv x \pmod{p^n}$, and that y is unique mod p^{n+1} (i.e., $\bar{y} \in \mathbb{Z}/p^{n+1}$ is unique).

Write $x^2 = a + bp^n$. We determine all $z \in \mathbb{Z}$ such that $y^2 \equiv a \pmod{p^{n+1}}$ with $y = x + zp^n$. Calculation shows that

$$y^2 = (x + zp^n)^2 \equiv a + (2xz + b)p^n \pmod{p^{n+1}}.$$

The condition on z is that $2xz \equiv -b \pmod{p}$. Such z exist because $p \nmid 2x$.

Moreover, z is uniquely determined mod p, which proves that y is unique mod p^{n+1}, as claimed.

We have produced a map from solutions $\bar{x}$ of the equation $X^2 = \bar{a} \in \mathbb{Z}/p^n$ to solutions $\bar{y}$ of the equation $X^2 = \bar{a} \in \mathbb{Z}/p^{n+1}$. It is inverse to the natural map $y + \mathbb{Z}/p^{n+1} \mapsto y + \mathbb{Z}/p^n$ on solutions, which is therefore a bijection. An induction quickly shows that the number of solutions of the equation $X^2 = \bar{a} \in \mathbb{Z}/p^n$ is the same for all $n \geq 1$. Since $\mathbb{Z}/p$ is a field, p is odd, and $p \nmid a$, this number is 0 or 2.

If x is one solution of the congruence $X^2 \equiv a \pmod{p^n}$, then $-x$ is the other.

ii. Let $n \geq 3$. The solutions of the congruence $X^2 \equiv a \pmod{2^n}$ come in pairs. If $x_1 \in \mathbb{Z}$ is a solution, then so is $x_2 = x_1 + 2^{n-1}$. We prove that there is a unique member x_i of each such pair for which there exists $y \in \mathbb{Z}$ such that $y^2 \equiv a \pmod{2^{n+1}}$ and $y \equiv x_i \pmod{2^n}$, and that there are exactly two such y mod 2^{n+1}.

Write $x_i^2 = a + b_i 2^n$. Computation shows that $y_i = x_i + z \cdot 2^n$ satisfies the congruence $y_i^2 \equiv x_i^2 \pmod{2^{n+1}}$ for all $z \in \mathbb{Z}$. Thus $y_i^2 \equiv a \pmod{2^{n+1}}$ if and only if b_i is even. But b_1 and $b_2 = b_1 + x_1 + 2^{n-2}$ have opposite parity because $n \geq 3$ and x_1 is odd, so exactly one of b_1, b_2 is even. This proves the existence part of the assertion. There are exactly two integers y mod 2^{n+1} as required because y_i is determined mod 2^{n+1} by the parity of z.

For each $n \geq 3$ let S_n denote the set of solutions of the equation $X^2 = \bar{a} \in \mathbb{Z}/2^n$. Let $g\colon S_{n+1} \to S_n$ be the natural map $x + \mathbb{Z}/2^{n+1} \mapsto x + \mathbb{Z}/2^n$. Both members of a pair of solutions in S_{n+1} as above have the same image under g, so g is two-to-one. Thus $|\text{im } g| = |S_{n+1}|/2$. Also $|\text{im } g| = |S_n|/2$, because the image of g contains exactly one member of each pair of elements in S_n. Therefore, the number of solutions of $X^2 = \bar{a} \in \mathbb{Z}/2^n$ is the same for all $n \geq 3$. Consideration of the case $n = 3$ shows that this number is 4 if $a \equiv 1 \pmod 8$ and is zero otherwise.

If x is one solution of the congruence $X^2 \equiv a \pmod{2^n}$, then the other three are $x + 2^{n-1}$, $-x$, and $-x + 2^{n-1}$. ■

Theorem 2.6. Let a and m be nonzero relatively prime integers. Write $m = 2^b d$ where d is odd. The following two conditions are equivalent.

1. $X^2 \equiv a \pmod m$ has a solution.
2. $X^2 \equiv a \pmod p$ has a solution for every odd prime divisor p of m and

$$a \equiv 1 \pmod 4 \quad \text{if } b = 2,$$

$$a \equiv 1 \pmod 8 \quad \text{if } b \geq 3.$$

Proof. Immediate consequence of Propositions 2.4 and 2.5. ■

Exercises

1. Find all integers x that $x \equiv 1 \pmod{17}$, $x \equiv 2 \pmod{19}$, and $x \equiv 3 \pmod{23}$.

2. Show that $x = 70a + 21b + 15c$ satisfies the three congruences $x \equiv a \pmod 3$, $x \equiv b \pmod 5$, and $x \equiv c \pmod 7$. Find a solution of a similar form for the system $x \equiv a \pmod 5$, $x \equiv b \pmod 7$, and $x \equiv c \pmod{11}$.

3. Find the remainder when $((92)^{397} + 100)^{199}$ is divided by $23 \cdot 199$.

4. Prove that $a^{80} \equiv 1 \pmod{561}$ for all integers a such that $\mathrm{GCD}(a, 561) = 1$.

5. Let $x, y, z \in \mathbb{Z}$ be such that $x^2 + y^2 = z^2$. Prove that $xyz \equiv 0 \pmod{60}$.

6. Let N be a positive integer. Prove that there exist N consecutive positive integers each of which is divisible by the tenth power of an integer greater than 1.

7. Let m_1, m_2 be positive integers and let a_1, $a_1 \in \mathbb{Z}$. Prove that there exists $x \in \mathbb{Z}$ such that $x \equiv a_1 \pmod{m_1}$ and $x \equiv a_2 \pmod{m_2}$ if and only if $a_1 \equiv a_2 \pmod{\mathrm{GCD}(m_1, m_2)}$.

8. Prove that the congruence $X^2 + 2X + 2 \equiv 0 \pmod{5^n}$ has a solution for every positive integer n.

9. Let $f \in \mathbb{Z}[X]$ and let p be a prime number.

i. Prove that $f(a + tp^n) \equiv f(a) + tp^n f'(a) \pmod{p^{n+1}}$ for all $a, t \in \mathbb{Z}$ and $n \geq 1 \in \mathbb{Z}$.

ii. Let $a \in \mathbb{Z}$ be such that $f(a) \equiv 0 \pmod p$ and $f'(a) \not\equiv 0 \pmod p$. Prove that for every $n \geq 1$ there exists $b \in \mathbb{Z}$ such that $b \equiv a \pmod p$ and $f(b) \equiv 0 \pmod{p^n}$. Prove moreover that such b is uniquely determined mod p^n.

iii. Suppose that the two congruence equations $f(X) \equiv 0 \pmod p$ and $f'(X) \equiv 0 \pmod p$ have no solutions in common. Prove that $|\{\bar{b} \in \mathbb{Z}/p^n | f(\bar{b}) = 0 \in \mathbb{Z}/p^n\}|$ is the same number for all integers $n \geq 1$.

10. Is Proposition 2.5 a special case of the assertion of Exercise 9iii?

11. Let p be a prime number and let $a = dp^r$ where $r \geq 1$ and $p \nmid d$. Let $n > r$. Prove that

$$\left|\{x \in \mathbb{Z}/p^n | x^2 = \bar{a}\}\right| = \begin{cases} 0 & \text{if } r \text{ is odd} \\ p^{r/2} \cdot \left|\{y \in \mathbb{Z}/p^{n-r} | y^2 = \bar{d}\}\right| & \text{if } r \text{ is even} \end{cases}.$$

3. The Legendre Symbol

We begin the theory of quadratic congruences with some classical terminology.

***Definition* 3.1.** The *Legendre symbol* (a/p) is defined for an odd prime number p and an integer a such that $\mathrm{GCD}(a, p) = 1$ by the equation

$$\left(\frac{a}{p}\right) = \begin{cases} +1 & \text{if the congruence } X^2 \equiv a \pmod{p} \text{ has a solution} \\ -1 & \text{if the congruence } X^2 \equiv a \pmod{p} \text{ has no solution} \end{cases}.$$

The integer a is said to be a *quadratic residue* mod p iff $(a/p) = +1$ and a *quadratic nonresidue* mod p iff $(a/p) = -1$.

Much of this chapter will be concerned with the problem of evaluating the Legendre symbol.

It is clear from Definition 3.1 that $(a/p) = (b/p)$ if $a \equiv b \pmod{p}$. Thus (y/p) can be defined for $y \in U_p$ by the equation $(y/p) = (a/p)$ where $y = \bar{a}$ with $a \in \mathbb{Z}$. Then $(y/p) = \pm 1$ for $y \in U_p$ depending on whether the equation $X^2 = y$ has a root in the field $\mathbb{Z}/p$. We will extend the words *quadratic residue* and *nonresidue* mod p to the elements of U_p in the obvious way.

The elementary properties of the Legendre symbol are collected in the next proposition.

Proposition 3.2. Let p be an odd prime number.

i. There exist exactly $(p-1)/2$ quadratic residues and $(p-1)/2$ quadratic nonresidues mod p in U_p.

ii. $a^{(p-1)/2} \equiv (a/p) \pmod{p}$ for all $a \in \mathbb{Z}$ such that $p \nmid a$.

iii. $(ab/p) = (a/p)(b/p)$ for all $a, b \in \mathbb{Z}$ such that $p \nmid ab$.

iv. The function $(\cdot/p)\colon U_p \to \{\pm 1\}$ is a surjective group homomorphism from U_p to the multiplicative group $\{\pm 1\}$.

Proof. i. The set of quadratic residues in U_p is the image of the homomorphism $sq\colon U_p \mapsto U_p$, $sq(y) = y^2$. The kernel of sq consists of the two roots $+1$ and -1 of the equation $X^2 - 1 = 0$ in $\mathbb{Z}/p$. We can compute: $|\mathrm{im}(sq)| = |U_p|/|\ker(sq)| = (p-1)/2$.

ii. By Fermat's Little Theorem 2.3 of Chapter 2, every element of U_p satisfies the equation

$$(X^{(p-1)/2} - 1)(X^{(p-1)/2} + 1) = X^{p-1} - 1 = 0 \in \mathbb{Z}/p.$$

Thus every element of U_p satisfies one of the two equations: $X^{(p-1)/2} - 1 = 0$ or $X^{(p-1)/2} + 1 = 0$.

If $y \in U_p$ is a quadratic residue, then $y = x^2$ for some $x \in U_p$, whence $y^{(p-1)/2} = x^{p-1} = 1$ in $\mathbb{Z}/p$. Thus all $(p-1)/2$ quadratic residues in U_p

satisfy the equation $X^{(p-1)/2} - 1 = 0$, which can have at most $(p-1)/2$ roots in $\mathbb{Z}/p$ by Lemma 2.6 of Chapter 2. The quadratic nonresidues in U_p must therefore all satisfy the other equation, $X^{(p-1)/2} + 1 = 0$.

We have proved that $y^{(p-1)/2} = (y/p) \in \mathbb{Z}/p$ for all $y \in U_p$, from which the desired result follows.

iii. Compute $(ab/p) \equiv (ab)^{(p-1)/2} = a^{(p-1)/2} b^{(p-1)/2} = (a/p)(b/p) \pmod{p}$. Since $-1 \not\equiv 1 \pmod{p}$, we can conclude that $(ab/p) = (a/p)(b/p)$.

iv. Immediate from iii and i. ∎

The multiplicative law Proposition 3.2iii is the key property of the Legendre symbol. It suggests the historical line of investigation, which we follow.

Let p be an odd prime and let $a \in \mathbb{Z}$ be relatively prime to p. We can factor a, $a = \pm 2^m \prod_i q_i^{n_i}$, where the q_i are odd primes distinct from p. Then Proposition 3.2iii shows that

$$\left(\frac{a}{p}\right) = \left(\frac{\pm 1}{p}\right)\left(\frac{2}{p}\right)^m \prod_i \left(\frac{q_i}{p}\right)^{n_i}.$$

We are led to focus our attention on the symbols $(-1/p)$, $(2/p)$, and (q/p) for distinct odd primes p, q, from which all other Legendre symbols may be computed.

Actually, we do not even need the symbols $(-1/p)$ and $(2/p)$. The equations $(-1/p) = (2p - 1/p)$ and $(2/p) = (p + 2/p)$ show that they may be computed from knowledge of (q/p) for all prime divisors q of the positive odd integers $2p - 1$ and $p + 2$. Nevertheless, it is convenient to state:

Proposition 3.3. Supplement to the Law of Quadratic Reciprocity. Let p be an odd prime number.

i. $$\left(\frac{-1}{p}\right) = (-1)^{(p-1)/2} = \begin{cases} 1 & \text{if } p \equiv 1 \pmod{4} \\ -1 & \text{if } p \equiv 3 \pmod{4} \end{cases}.$$

ii. $$\left(\frac{2}{p}\right) = (-1)^{(p^2-1)/8} = \begin{cases} 1 & \text{if } p \equiv \pm 1 \pmod{8} \\ -1 & \text{if } p \equiv \pm 3 \pmod{8} \end{cases}.$$

Proof. i. This is just a reformulation of Lemma 1.4 of Chapter 2, on which we based the study of the Diophantine equation $X^2 + Y^2 = n$. It also follows immediately from Proposition 3.2ii.

ii. We first prove that $(2/p) = 1$ implies $p \equiv \pm 1 \pmod{8}$. This will be done by induction on p.

Since $(2/3) = (2/5) = -1$ and $(2/7) = +1$, the induction begins with $p = 7$, which is in fact congruent to $-1 \bmod 8$ as required.

Now let $p > 7$ be an odd prime such that $(2/p) = 1$, and suppose that $q \equiv \pm 1 \pmod{8}$ for all primes $q < p$ for which $(2/q) = 1$. Since $(2/p) = 1$, we can choose integers u and a, $0 < a < p$, such that $a^2 = 2 + up$. After replacing a by $p - a$ if necessary, we may take a to be odd. Clearly then u is odd and $1 \le u < p$. The inductive hypothesis applies to every prime divisor q of u since $a^2 \equiv 2 \pmod{q}$, and so every such q is congruent to $\pm 1 \pmod{8}$. Because u is a product of these qs, we conclude that u is also congruent to $\pm 1 \pmod{8}$. Finally, we compute $1 \equiv a^2 = 2 + up \equiv 2 \pm p \pmod{8}$, which shows that $p \equiv \pm 1 \pmod{8}$. The induction is complete.

A completely similar induction proves that $(-2/p) = 1$ implies $p \equiv 1$ or $3 \pmod{8}$. (Details are left as an exercise.)

Now let p be an odd prime number.

If $p \equiv \pm 3 \pmod{8}$, then the first induction proves that $(2/p) = -1$.

If $p \equiv -1 \pmod{8}$, then the second induction shows that $(-2/p) = -1$ and Proposition 3.3i shows that $(-1/p) = -1$. By the multiplicative property, $(2/p) = (-1/p)(-2/p) = 1$.

If $p \equiv 1 \pmod{8}$, let $z = y^{(p-1)/8}$ where y is a primitive root mod p. Then $z^4 \equiv -1 \pmod{p}$, from which follows $(z^3 - z)^2 = z^2(z^4 + 1) - 2z^4 \equiv 2 \pmod{p}$. Thus $(2/p) = 1$.

We have evaluated $(2/p)$ in all cases. ■

We are left to face the basic fact in the theory of quadratic congruences, the relation between (p/q) and (q/p) for distinct odd prime numbers p and q that is described by Proposition 1.2. We next give three traditional formulations of this relation, now known as the Law of Quadratic Reciprocity. That Proposition 1.2 is equivalent to Proposition 3.4iii, which is the statement that Gauss preferred, follows from Proposition 3.3i.

Proposition 3.4. The Law of Quadratic Reciprocity: Three Formulations of Proposition 1.2

i.
$$\left(\frac{q}{p}\right) = (-1)^{(p-1)/2\cdot(q-1)/2}\cdot\left(\frac{p}{q}\right) = \begin{cases} \left(\frac{p}{q}\right) & \text{if } p \equiv 1 \text{ or } q \equiv 1 \pmod{4} \\ -\left(\frac{p}{q}\right) & \text{if } p \equiv q \equiv 3 \pmod{4} \end{cases}$$

ii.
$$\left(\frac{q}{p}\right)\left(\frac{p}{q}\right) = (-1)^{(p-1)/2\cdot(q-1)/2}.$$

iii.
$$\left(\frac{p^*}{q}\right) = \left(\frac{q}{p}\right) \quad \text{where } p^* = (-1)^{(p-1)/2}p.$$

Four quite different proofs of the Law of Quadratic Reciprocity will be the subject of the next four sections of this chapter.

The importance of quadratic reciprocity will become clear only upon its application to the study of integral binary quadratic forms. We can, however, begin to discuss its meaning now.

A good understanding of the Legendre symbol (a/p) viewed as a function of a for fixed p can be arrived at through Proposition 3.2. The Law of Quadratic Reciprocity enables us to say something interesting about (a/p) as a function of p for fixed a. As Euler discovered empirically, the value of (a/p) depends upon p only through the congruence class of $p \bmod 4a$. This is illustrated nicely by the cases $a = -1$ and $a = 2$ of Proposition 3.3.

Proposition 3.5. Let $a \in \mathbb{Z}$, and let p and q be odd prime numbers not dividing a such that $p \equiv q \pmod{4a}$.

Then $(a/p) = (a/q)$.

Proof. Write $a = \pm 2^m \prod_i r_i$, where the r_i are odd primes distinct from p and q. We will prove: (a) that $(-1/p) = (-1/q)$, (b) that $(2/p) = (2/q)$ if a is even, and (c) that $(r_i/p) = (r_i/q)$ for all i. The proposition will then follow directly from an application of the multiplicative property Proposition 3.2iii of the Legendre symbol.

By hypothesis $p \equiv q \pmod 4$, and so (a) is a consequence of Proposition 3.3i. Similarly, if a is even, then $p \equiv q \pmod 8$ and so (b) follows from Proposition 3.3ii.

Finally, the congruence $p \equiv q \pmod{r_i}$ implies that $(p/r_i) = (q/r_i)$. If $p \equiv q \equiv 1 \pmod 4$, then by Proposition 3.4i we have $(r_i/p) = (p/r_i) = (q/r_i) = (r_i/q)$. If $p \equiv q \equiv 3 \pmod 4$, then Proposition 3.4i shows that $(r_i/p) = (-1)^{(r_i-1)/2}(p/r_i) = (-1)^{(r_i-1)/2}(q/r_i) = (r_i/q)$. Thus (c) is proved. ■

Gauss's statement Proposition 3.4iii prompts the next definition and lemma, which will be used occasionally in subsequent sections.

Definition 3.6. We define m^* for an odd integer m by the formula $m^* = (-1)^{(m-1)/2} m$.

Lemma 3.7

i. $|m|^* = m^*$ for all odd integers m.

ii. $(\prod_i m_i)^* = \prod_i m_i^*$ for odd integers m_i.

iii. $(-1)^{(m-1)/2} = \operatorname{sign}(m \cdot m^*)$ for all odd integers m.

iv. $(-1)^{(m-1)/2}$ is multiplicative in odd integers m. In other words, $(-1)^{(m_1 m_2 - 1)/2} = (-1)^{(m_1-1)/2} \cdot (-1)^{(m_2-1)/2}$ for all pairs of odd integers m_1 and m_2.

Proof. Remark first that m^* is the unique integer such that $|m^*| = |m|$ and $m^* \equiv 1 \pmod 4$. Parts i and ii are immediate consequences. To prove iii, note that $m = m^*$ if $m \equiv 1 \pmod 4$ and that $m = -m^*$ if $m \equiv 3 \pmod 4$. Part iv follows from ii and iii and the multiplicativity of the sign function. ∎

Exercises

1. Let p be an odd prime and let y be a primitive root mod p. Prove that $(y/p) = -1$.

2. Let p be an odd prime number. Prove that the smallest positive quadratic nonresidue mod p is prime.

3. Show that for all primes $p \geq 7$ there is an integer n, $1 \leq n \leq 9$, such that $(n/p) = ((n+1)/p) = 1$. (*Suggestion:* Consider separately the three cases $(2/p) = 1$, $(5/p) = 1$, and $(2/p) = (5/p) = -1$.)

4. Complete the proof of Proposition 3.3ii by showing that $(-2/p) = 1$ for an odd prime p implies that $p \equiv 1$ or $3 \pmod 8$.

5. Let $p \neq 5$ be an odd prime number such that $(5/p) = 1$. Show by induction on p that $p \equiv \pm 1 \pmod 5$. (*Sketch:* Write $a^2 = 5 + up$, where $0 < a < p$ and a is even. If $5 \nmid u$, show that $u \equiv \pm 1 \pmod 5$ by applying the inductive hypothesis to all prime divisors of u. If $5|u$, show similarly that $u/5 \equiv \pm 1 \pmod 5$.)

6. (Lagrange's proof that $(5/p) = +1$ for every odd prime $p \equiv 4 \pmod 5$.) We say that two polynomials $f = \Sigma a_i X^i$ and $g = \Sigma b_i X^i$ in $\mathbb{Z}[X]$ are congruent mod n iff $a_i \equiv b_i \pmod n$ for all i. Let p be an odd prime number and let $b \in \mathbb{Z}$ be a quadratic nonresidue mod p.

   i. Prove that

$$f = \frac{(X + \sqrt{b})^{p+1} - (X - \sqrt{b})^{p+1}}{\sqrt{b}} \in \mathbb{Z}[X].$$

   Show that

$$\deg(f) = p \text{ and that } f \equiv 2(p+1)(X^p - X) \pmod p.$$

   ii. Let $e \in \mathbb{Z}$ be a positive divisor of $p + 1$, and let

$$g = \frac{(X + \sqrt{b})^{e} - (X - \sqrt{b})^{e}}{\sqrt{b}}.$$

Show that $g \in \mathbb{Z}[X]$. Prove that there exists $h \in \mathbb{Z}[X]$ such that $f = gh$. Conclude that if $e > 1$ there exists $a \in \mathbb{Z}$ such that $g(a) \equiv 0 \pmod{p}$.

iii. Suppose that $p \equiv 4 \pmod{5}$. Let $e = 5$ in ii and verify that $g = 10X^4 + 20bX^2 + 2b^2$. Conclude that $(5/p) = +1$.

7. Evaluate $(211/317)$.

8. Prove that the congruence $X^2 \equiv 3 \pmod{p}$ has a solution for a prime $p > 3$ if and only if $p \equiv \pm 1 \pmod{12}$.

9. Find all odd primes $p \neq 5$ such that $(5/p) = 1$.

10. For which primes p does $X^2 \equiv 7 \pmod{p}$ have a solution? Same question for the congruence $X^2 \equiv 13 \pmod{p}$.

11. Let $f(X) = rX + s \in \mathbb{Q}[X]$ be such that $f(p) \in \mathbb{Z}$ and $(-1/p) = (-1)^{f(p)}$ for all odd primes p. Show that $f(X) = a(X-1)/2 + b$ for some $a, b \in \mathbb{Z}$ with a odd and b even.

12. Prove directly that $(-1)^{(m-1)/2}$ is multiplicative in odd integers m.

13. Let $f(X) = rX^2 + sX + t \in \mathbb{Q}[X]$ be such that $f(p) \in \mathbb{Z}$ and $(2/p) = (-1)^{f(p)}$ for all odd primes p. Show that $f(X) = a(X^2 - 1)/8 + bX + c$, where $a, b, c \in \mathbb{Z}$ with a odd and $b \equiv c \pmod{2}$.

14. Show that 503 is a prime divisor of $2^{251} - 1$.

15. Let $p \equiv \pm 1 \pmod{8}$ be prime. Show that 2 is not a primitive root mod p.

16. Show that 2 is a primitive root mod 347.

17. Determine whether -2 is a primitive root mod 359.

18. Let p be an odd prime such that $q = 2p + 1$ is also prime. Prove that 2 is a primitive root mod q if $p \equiv 1 \pmod{4}$ and that -2 is a primitive root mod q if $p \equiv 3 \pmod{4}$.

19. Let p be a prime such that $q = 2p + 1$ is also prime and let $a \in \mathbb{Z}$. Prove that a is a primitive root mod q if and only if $a \not\equiv 0$ or $-1 \pmod{q}$ and $(a/q) = -1$.

20. Let p be a prime such that $q = 4p + 1$ is also prime. Prove that 2 is a primitive root mod q. Find the four smallest primes q to which this exercise applies.

21. For which primes p with $q = 2p + 1$ prime is 5 a primitive root mod q?

22. Show that 3 is a primitive root mod every prime of the form $2^n + 1$, $n \geq 2$.

23. Prove that there is an infinite number of primes of each of the two forms $8n+3$, $8n+7$. (*Hint:* Let P be a product of primes of the desired type. Consider prime divisors of P^2+2, $8P^2-1$.)

24. Prove that there is an infinite number of primes of the form $5n-1$. Hence prove that there is an infinite number of primes with final digit 9.

25. Let $p \equiv -1 \pmod 8$ be prime and let $x = 2^{(p+1)/4}$. Prove that $x^2 \equiv 2 \pmod p$.

26. Let p be an odd prime and let $a \in \mathbb{Z}$ where $p \nmid a$ and $(a/p) = 1$.

i. Suppose that $p \equiv 3 \pmod 4$. Show that $x^2 \equiv a \pmod p$, where $x = a^{(p+1)/4}$.

ii. Suppose that $p \equiv 5 \pmod 8$. Prove that $x^2 \equiv \pm a \pmod p$ and $y^2 \equiv -x^2 \pmod p$, where $x = a^{(p+3)/8}$ and $y = 2^{(p-1)/4}x$.

27. Let $p \equiv 1 \pmod 4$ be prime and write $p = a^2 + b^2$, where $a, b \in \mathbb{Z}$ and a is odd. Prove that $(a/p) = 1$ and $(b/p) = (-1)^{(p-1)/4}$.

28. Let $p \neq 3$ be an odd prime. Express $(3/p)$ as a product of Legendre symbols (q/p) with primes $q > 3$.

29. Let a be a positive integer and let p and q be odd prime numbers not dividing a. Prove that if $p \equiv -q \pmod{4a}$, then $(a/p) = (a/q)$.

30. Deduce the Law of Quadratic Reciprocity from Proposition 3.5 and the assertion of Exercise 29.

31. Let p be an odd prime. Let

$$RN = \left|\left\{ n \in \mathbb{Z} \,\middle|\, 1 \le n \le p-2, \left(\frac{n}{p}\right) = 1, \left(\frac{n+1}{p}\right) = -1 \right\}\right|.$$

Similarly let RR, NR, NN be the number of integers n such that $1 \le n \le p-2$ with values for (n/p) and $((n+1)/p)$ specified by the mnemonic "R for quadratic *R*esidue and N for quadratic *N*onresidue mod p."

i. Show that

$$RR + RN = \begin{cases} (p-1)/2 - 1 & \text{if } p \equiv 1 \pmod 4 \\ (p-1)/2 & \text{if } p \equiv 3 \pmod 4 \end{cases}.$$

Evaluate similarly $NR + NN$, $RR + NR$, and $RN + NN$.

ii. Show that $RR + NN - RN - NR = -1$. (*Hint:* Show that $\sum_{n=1}^{p-2} (n(n+1)/p) = \sum_{m=1}^{p-2} (1(1+m)/p) = -1$.)

iii. Prove that

$$(RR, RN, NR, NN) = \begin{cases} (x-1, x, x, x) & \text{if } p = 4x+1 \equiv 1 \pmod 4 \\ (x, x+1, x, x) & \text{if } p = 4x+3 \equiv 3 \pmod 4 \end{cases}.$$

4. The First Proof

Gauss completed his first proof of the Law of Quadratic Reciprocity when he was 19 years old. The proof is an induction. We begin our presentation of that proof with a lemma that is crucial at the inductive step. The proof of the lemma is the most amazing proof in this book. Gauss could be extraordinarily clever.

Lemma 4.1. Let q be a prime congruent to 1 mod 4. Then there exists an odd prime $p < q$ such that $(q/p) = -1$.

Proof. This is easy if $q \equiv 5 \pmod 8$. Then $(q+1)/2 \equiv 3 \pmod 4$, so there is a prime divisor p of $(q+1)/2$ that is congruent to 3 mod 4. Clearly $p < q$. Since $q \equiv -1 \pmod p$ we can compute, using Proposition 3.3i, that $(q/p) = (-1/p) = -1$.

Suppose that $q \equiv 1 \pmod 8$. Suppose that $m \in \mathbb{Z}$ is such that $1 < 2m+1 < q$ and $(q/p) = +1$ for all odd primes $p \leq 2m+1$. By Theorem 2.6, there exists $x > m$ such that $x^2 \equiv q \pmod{(2m+1)!}$. We compute

$$\begin{aligned}(q-1^2)(q-2^2)\cdots(q-m^2) &\equiv (x^2-1^2)(x^2-2^2)\cdots(x^2-m^2) \\ &= (2m+1)!\binom{x+m}{2m+1}\Big/x \pmod{(2m+1)!}.\end{aligned}$$

Since the binomial coefficient is an integer and $\mathrm{GCD}(x,(2m+1)!) = 1$, we can conclude by Lemma 3.6 of Chapter 1 that

$$\begin{aligned}&\frac{(q-1^2)(q-2^2)\cdots(q-m^2)}{(2m+1)!} \\ &= \frac{1}{m+1}\frac{q-1^2}{(m+1)^2-1^2}\cdots\frac{q-m^2}{(m+1)^2-m^2} \in \mathbb{Z}.\end{aligned}$$

This is impossible if $m^2 < q < (m+1)^2$, because then every factor in the last product is a fraction between 0 and 1. This remark applies to $m = [\sqrt{q}]$. Since $2[\sqrt{q}]+1 < q$, there must exist an odd prime $p \leq 2[\sqrt{q}]+1$ such that $(q/p) = -1$. ■

Proof of Proposition 1.2, The Law of Quadratic Reciprocity. We will prove Proposition 3.4iii: $(p^*/q) = (q/p)$ for all pairs of distinct odd primes p, q. By Proposition 3.3i the roles of p and q are symmetric, so we assume that $p < q$. The proof will be by induction on q, the maximum of the two primes. The induction begins with the trivial case $(3^*/5) = (5/3) = -1$.

Let $q > 5$ and suppose that $(a^*/b) = (b/a)$ for all pairs of distinct odd primes a and b both less than q. Let $p < q$. We divide the analysis into three cases.

***Case 1.* $(p^*/q) = 1$.** We are to prove that $(q/p) = 1$.

Choose integers u and a, $0 < a < q$, such that $a^2 = p^* + uq$. Replacing a by $q - a$ if necessary, we take a to be even. Then u is odd and the inequality $-q < a^2 - p^* \leq (q-1)^2 + p < q^2 - q$ shows that $1 \leq u < q$.

If $p \nmid u$, then $p \nmid a$. The congruence $a^2 \equiv uq \pmod{p}$ implies that $(q/p) = (u/p)$. We must show that $(u/p) = 1$. Write $u = \prod_i p_i$ where the p_i are (not necessarily distinct) odd primes less than q and different from p. Since $a^2 \equiv p^* \pmod{p_i}$, we find that $(p^*/p_i) = 1$. By the induction hypothesis, $(p_i/p) = 1$. Thus $(u/p) = \prod_i (p_i/p) = 1$.

If $p|u$, then $p|a$. Let $A = a/p$ and $U = u/p$. We then have $pA^2 = (-1)^{(p-1)/2} + Uq$, where $p \nmid U$. We can proceed much as in the previous paragraph. The congruence $(-1)^{(p+1)/2} \equiv Uq \pmod{p}$ shows that $(Uq/p) = (-1)^{(p-1)/2 \cdot (p+1)/2} = 1$ and hence that $(q/p) = (U/p)$. We evaluate (U/p). Write $U = \prod_i p_i$, with p_i prime. The congruence $p^*A^2 \equiv 1 \pmod{p_i}$ shows that $(p^*/p_i) = 1$. The induction hypothesis applies to show that $(p_i/p) = 1$. Therefore, $(U/p) = \prod_i (p_i/p) = 1$, as desired.

***Case 2.* $q \equiv 3 \pmod{4}$ *and* $(p^*/q) = -1$.** We are to prove that $(q/p) = -1$.

The hypotheses imply that $(-p^*/q) = 1$. Let u and a be integers with a even and $0 < a < q$ such that $a^2 = -p^* + uq$. Then, as in Case 1, u is odd and $1 \leq u < q$.

If $p \nmid u$, then $p \nmid a$ and $(q/p) = (u/p)$. We must show that $(u/p) = -1$. Write $u = \prod_i p_i$ where the p_i are prime. Then $(-p^*/p_i) = 1$, which yields $(p^*/p_i) = (-1)^{(p_i-1)/2}$. The induction hypothesis implies that $(p_i/p) = (p^*/p_i)$. Hence $(u/p) = \prod_i (p_i/p) = (-1)^{(u-1)/2}$, where we have used Lemma 3.7iv. Since $u \equiv -1 \pmod{4}$, we can conclude that $(u/p) = -1$.

If $p|u$, then $p|a$. Let $A = a/p$ and $U = u/p$. We then have $pA^2 = (-1)^{(p+1)/2} + Uq$, where $p \nmid U$. Since $(Uq/p) = ((-1)^{(p-1)/2}/p) = (-1)^{(p-1)/2}$, we find that we are to show that $(U/p) = (-1)^{(p+1)/2}$. Write $U = \prod_i p_i$, with p_i prime. Applying the induction hypothesis, we compute $(p_i/p) = (p^*/p_i) = (-1/p_i) = (-1)^{(p_i-1)/2}$. Thus $(U/p) = \prod_i (p_i/p) = (-1)^{(U-1)/2}$. Noting that $U \equiv -p \pmod{4}$, we conclude that $(U/p) = (-1)^{(p+1)/2}$ as desired.

Case 3. $q \equiv 1 \pmod 4$ ***and*** $(p^*/q) = -1$. We are to prove that $(q/p) = -1$.

We must work a little to push Case 3 into a form similar to that of the first two cases.

By Lemma 4.1 there exists an odd prime p' less than q such that $(q/p') = -1$. If $p = p'$ we are done, so assume that $p \neq p'$. If $(p'/q) = 1$, then by Case 1 we would have $(q/p') = 1$, a contradiction. Therefore $(p'/q) = -1$. We are to prove that $(q/p)(q/p') = 1$. Suppose more generally that p and p' are any two distinct odd primes less than q such that $(pp'/q) = 1$. We will show that $(q/p)(q/p') = 1$.

Since $(pp'/q) = 1$, we can find an integer u and an even integer a, $0 < a < q$, such that $a^2 = pp' + uq$. Then u is odd and $1 \le |u| < q$. There are three cases to be considered.

If $p \nmid u$ and $p' \nmid u$, then $\mathrm{GCD}(a, pp') = 1$. The congruence $a^2 \equiv uq \pmod{pp'}$ shows that $(q/p) = (u/p)$ and $(q/p') = (u/p')$, so we must prove that $(u/p)(u/p') = 1$. Write $|u| = \prod_i p_i$ with p_i prime. The equation $(pp'/p_i) = 1$ becomes, on application of the induction hypothesis to (p/p_i) and (p'/p_i), the equation $(p_i^*/p)(p_i^*/p') = 1$. Thus

$$\begin{aligned} 1 &= \prod_i (p_i^*/p)(p_i^*/p') = (|u|^*/p)(|u|^*/p') = (u^*/p)(u^*/p') \\ &= (-1)^{(pp'-1)/2 \cdot (u-1)/2}(u/p)(u/p'), \end{aligned}$$

where we have used Lemma 3.7ii, i, and iv. Since $u \equiv -pp' \pmod 4$, we conclude that $(u/p)(u/p') = 1$.

If $p | u$ and $p' \nmid u$, then $p | a$. Let $A = a/p$ and $U = u/p$. Then $pA^2 = p' + Uq$ and $\mathrm{GCD}(pp', U) = 1$. Clearly $(Uq/p) = (-p'/p)$ and $(Uq/p') = (p/p')$. The induction hypothesis applies to (p'/p) and (p/p'), so our goal is to show that

$$\begin{aligned} (U/p)(U/p') &= (-p'/p)(p/p') = (-1)^{(p-1)/2 + ((p-1)/2 \cdot (p'-1)/2)} \\ &= (-1)^{(p-1)/2 \cdot (p'+1)/2}. \end{aligned}$$

Write $|U| = \prod_i p_i$, with p_i prime. Since $pA^2 \equiv p' \pmod{p_i}$, we have $(p/p_i)(p'/p_i) = 1$. Another application of the induction hypothesis yields

$$\begin{aligned} 1 &= \prod_i (p_i^*/p)(p_i^*/p') = (U^*/p)(U^*/p') \\ &= (-1)^{(pp'-1)/2 \cdot (U-1)/2}(U/p)(U/p'). \end{aligned}$$

Since $U \equiv -p' \pmod 4$, we have

$$(U/p)(U/p') = (-1)^{(pp'-1)/2\cdot(-p'-1)/2} = (-1)^{((p-1)/2+(p'-1)/2)(p'+1)/2}$$
$$= (-1)^{(p-1)/2\cdot(p'+1)/2},$$

as required.

Finally, if $pp'|u$, then $pp'|a$. Let $A = a/pp'$ and $U = u/pp'$. We then have $pp'A^2 = 1 + Uq$, where GCD$(pp', U) = 1$. Since $(Uq/p)(Uq/p') = (-1/p)(-1/p') = (-1)^{(pp'-1)/2}$, we must prove that $(U/p)(U/p') = (-1)^{(pp'-1)/2}$. Let $|U| = \prod_i p_i$ with p_i prime. Using the induction hypothesis we have $1 = (pp'/p_i) = (p_i^*/p)(p_i^*/p')$. Multiplying over i we get

$$1 = (U^*/p)(U^*/p') = (-1)^{(pp'-1)/2\cdot(U-1)/2}(U/p)(U/p')$$
$$= (-1)^{(pp'-1)/2}(U/p)(U/p'),$$

where the last equality holds because $U \equiv -1 \pmod 4$.

The first proof of quadratic reciprocity is now complete. ■

5. The Gauss Lemma

Gauss found a simple combinatorial evaluation of the Legendre symbol, now known as the Gauss Lemma, that was the basis of his third proof of quadratic reciprocity.

Lemma 5.1. Gauss's Lemma. Let p be an odd prime number and let $a \in \mathbb{Z}$ with $p \nmid a$. Let

$$X = \begin{pmatrix} 1 & 2 & \cdots & \dfrac{p-1}{2} \\ -1 & -2 & \cdots & -\dfrac{p-1}{2} \end{pmatrix}.$$

Let $Y = (a, 2a, \ldots, ((p-1)/2)a)$. Let N equal the number of elements of the sequence Y that are congruent mod p to some integer of the second row of the matrix X.

Then $(a/p) = (-1)^N$.

Proof. First observe that the elements of the matrix X form a complete set of representatives for the nonzero congruence classes mod p. Indeed, let j, k be elements of X. Then $|j-k| \leq p-1$, which shows that distinct elements of X cannot be congruent mod p. Since there are only $p-1$ nonzero congruence classes mod p, the $p-1$ elements of X must exhaust them.

Therefore, every element of Y is congruent mod p to exactly one element of X. For $j = 1, 2, \ldots, (p-1)/2$, we can hence define $\epsilon_j \in \{\pm 1\}$ and $\sigma(j) \in \{1, 2, \ldots, (p-1)/2\}$ by the condition $ja \equiv \epsilon_j \sigma(j) \pmod{p}$.

Next note that σ is a permutation of the set of integers from 1 to $(p-1)/2$. It will suffice to prove that σ is injective, which just means that distinct elements of Y are congruent mod p to elements of X from distinct columns of X. If ja, ka in Y are congruent to elements from X lying in the same column, then either $ja \equiv ka$ or $ja \equiv -ka \pmod{p}$. Thus $(j+k)a \equiv 0 \pmod{p}$ or $(j-k)a \equiv 0 \pmod{p}$, from which it follows that $p | j+k$ or $p | j-k$. But $0 < j+k < p$ and $|j-k| \leq p-1$, so we must have $j = k$. The elements ja, ka are not distinct, which proves what we wanted.

We can summarize what we have proved so far by saying that there is exactly one element in each column of X that is congruent mod p to an element of Y.

Since σ is a permutation, we have $\prod_{j=1}^{(p-1)/2} \sigma(j) = ((p-1)/2)!$.

The number N of the lemma is precisely the number of $j = 1, 2, \ldots, (p-1)/2$ such that $\epsilon_j = -1$. Thus $\prod_{j=1}^{(p-1)/2} \epsilon_j = (-1)^N$.

Finally we can calculate

$$a^{(p-1)/2} \cdot \left(\frac{p-1}{2}\right)! = \prod_{j=1}^{(p-1)/2} ja \equiv \prod_{j=1}^{(p-1)/2} \epsilon_j \sigma(j)$$

$$= (-1)^N \left(\frac{p-1}{2}\right)! \pmod{p}.$$

Because $((p-1)/2)!$ is relatively prime to p, it can be cancelled from the congruence, so we have $a^{(p-1)/2} \equiv (-1)^N \pmod{p}$. By Proposition 3.2ii we can conclude that $(a/p) \equiv (-1)^N \pmod{p}$, and the Gauss Lemma is proved. ∎

As a first application of the Gauss Lemma we present a second proof of the Supplement to the Law of Quadratic Reciprocity.

Proof of Proposition 3.3. i. $(-1/p) = (-1)^N$, where the N of the Gauss Lemma is trivially seen to equal $(p-1)/2$.

ii. To compute $(2/p)$ by the Gauss Lemma, we must count the number N of $j \in \{1, 2, \ldots, (p-1)/2\}$ such that $(p+1)/2 \leq 2j \leq p-1$. Clearly N equals the number of integers j such that $(p+1)/4 \leq j \leq (p-1)/2$.

If $p = 4x + 1 \equiv 1 \pmod 4$, then $(p+1)/4 = x + \frac{1}{2}$, $(p-1)/2 = 2x$, and so $N = x$. Thus

$$\left(\frac{2}{p}\right) = (-1)^x = \begin{cases} 1 & \text{if } x \text{ is even, i.e., } p \equiv 1 \pmod 8 \\ -1 & \text{if } x \text{ is odd, i.e., } p \equiv 5 \pmod 8 \end{cases}.$$

Similarly, if $p = 4x + 3 \equiv 3 \pmod 4$, then $N = x + 1$. Hence

$$\left(\frac{2}{p}\right) = (-1)^{x+1} = \begin{cases} -1 & \text{if } x \text{ is even, i.e., } p \equiv 3 \pmod 8 \\ 1 & \text{if } x \text{ is odd, i.e., } p \equiv 7 \pmod 8 \end{cases}. \qquad \blacksquare$$

We now give a second proof of the Law of Quadratic Reciprocity.

Proof of Proposition 1.2. Let p and q be distinct odd primes. We will prove the law in the formulation of Proposition 3.4ii: $(q/p)(p/q) = (-1)^{(p-1)/2\cdot(q-1)/2}$.

During the proof we will work with the set $W = \{(x, y) \in \mathbb{Z}^2 | 0 < x < (p+1)/2 \text{ and } 0 < y < (q+1)/2\}$. W has $(p-1)/2 \cdot (q-1)/2$ elements since $x \in \{1, 2, \ldots, (p-1)/2\}$ and $y \in \{1, 2, \ldots, (q-1)/2\}$.

To compute (q/p) by the Gauss Lemma, we must count the number of $x \in \{1, 2, \ldots, (p-1)/2\}$ such that there exists (necessarily unique) $j \in \{1, 2, \ldots, (p-1)/2\}$ with $qx \equiv -j \pmod p$. This congruence can be written as an equality $qx = -j + py$, where $y \in \mathbb{Z}$, and then

$$0 < y = \frac{qx + j}{p} \leq \left(q\frac{p-1}{2} + \frac{p-1}{2}\right)\Big/ p = \frac{q+1}{2}\,\frac{p-1}{p} < \frac{q+1}{2}.$$

The Gauss Lemma thus asserts that $(q/p) = (-1)^{|M|}$, where $M = \{(x, y) \in W | 0 < py - qx < (p+1)/2\}$.

Interchanging p and q and x and y we find similarly that $(p/q) = (-1)^{|N|}$, where $N = \{(x, y) \in W | -(q+1)/2 < py - qx < 0\}$.

Two more sets R and S next enter the picture: $R = \{(x, y) \in W | py - qx \geq (p+1)/2\}$ and $S = \{(x, y) \in W | py - qx \leq -(q+1)/2\}$.

The four subsets M, N, R, S of W are clearly pairwise disjoint, and their union is all of W because there can be no integral solutions of the equation $py - qx = 0$ such that $0 < x < (p+1)/2 < p$. Therefore

$$(-1)^{(p-1)/2\cdot(q-1)/2} = (-1)^{|W|} = (-1)^{|M|+|N|+|R|+|S|} = \left(\frac{q}{p}\right)\left(\frac{p}{q}\right)(-1)^{|R|+|S|}.$$

To complete the proof of quadratic reciprocity it remains only to show that $|R| + |S|$ is even. We will in fact prove that $|R| = |S|$.

Define ϕ: $W \mapsto W$ by the formula

$$\phi(x, y) = ((p+1)/2 - x, (q+1)/2 - y).$$

Clearly ϕ is a bijection on W. Writing $\phi(x, y) = (x', y')$, one computes that $py' - qx' + (q+1)/2 = qx - py + (p+1)/2$, from which it is readily seen that $(x, y) \in R$ if and only if $(x', y') \in S$. Thus ϕ is a bijection from R to S. ∎

Exercises

1. Evaluate $(3/p)$ for primes $p > 3$ by direct application of the Gauss Lemma.

2. Deduce Proposition 3.5 directly from the Gauss Lemma. (*Suggestion:* Begin with the case $a > 0$. Let $p = 4ak + r$. Show that $(a/p) = (-1)^N$ where N, the number of multiples of a in all the open intervals $((n - \frac{1}{2})p, np)$ for $n = 1, 2, \ldots, [a/2]$, has the same parity as the total number of integers in all the intervals $((n - \frac{1}{2})r/a, nr/a)$ for $n = 1, 2, \ldots, [a/2]$.)

6. Gauss Sums

Gauss's fourth proof of quadratic reciprocity rests on the evaluation of an important finite sum that now bears his name.

Definition 6.1. For odd positive integers n, set $S(n) = \sum_{k=0}^{n-1} e^{2\pi i k^2/n}$.

After years and great effort, Gauss was able to prove:

Theorem 6.2

$$S(n) = \begin{cases} \sqrt{n} & \text{if } n \equiv 1 \pmod 4 \\ i\sqrt{n} & \text{if } n \equiv 3 \pmod 4 \end{cases}.$$

We begin by isolating a useful lemma.

Lemma 6.3. Let a, n be nonzero integers such that $n \nmid a$. Then $\sum_{k=0}^{n-1} e^{2\pi i a k/n} = 0$.

Proof. This results from $\sum_{k=0}^{n-1} X^k = (X^n - 1)/(X - 1)$ applied to $X = e^{2\pi i a/n} \neq 1$. ∎

Proof of Proposition 1.2 from Theorem 6.2. Let p, q be distinct odd primes. Quadratic reciprocity follows immediately from Theorem 6.2 and the computation $S(pq) = (p/q)(q/p)S(p)S(q)$, which we now verify.

First derive an alternative expression for $S(p)$.

$$S(p) = 1 + \sum_{k=1}^{p-1} e^{2\pi i k^2/p} = 1 + \sum_{r=1}^{p-1}\left(1 + \left(\frac{r}{p}\right)\right)e^{2\pi i r/p}$$

$$= \sum_{r=1}^{p-1}\left(\frac{r}{p}\right)e^{2\pi i r/p}. \tag{6.4}$$

To see the second equality, note that the sequence $1^2, 2^2, \ldots, (p-1)^2$ contains each quadratic residue of p twice and contains no quadratic nonresidues. The third equality follows from Lemma 6.3.

Finally, we compute.

$$S(pq) = \sum_{k=0}^{p-1}\sum_{l=0}^{q-1} e^{2\pi i(kq+lp)^2/pq} \quad \text{(by Proposition 2.3)}$$

$$= \left(\sum_{k=0}^{p-1} e^{2\pi i k^2 q/p}\right)\left(\sum_{l=0}^{q-1} e^{2\pi i l^2 p/q}\right)$$

$$= \left(1 + 2\sum_{\substack{r=1\\(r/p)=(q/p)}}^{p-1} e^{2\pi i r/p)}\right)\left(1 + 2\sum_{\substack{s=1\\(s/q)=(p/q)}}^{q-1} e^{2\pi i s/q}\right)$$

$$= \left(1 + \sum_{r=1}^{p-1}\left(1 + \left(\frac{q}{p}\right)\left(\frac{r}{p}\right)\right)e^{2\pi i r/p}\right)\left(1 + \sum_{s=1}^{q-1}\left(1 + \left(\frac{p}{q}\right)\left(\frac{s}{q}\right)\right)e^{2\pi i s/q}\right)$$

$$= \left(\frac{q}{p}\right)\sum_{r=1}^{p-1}\left(\frac{r}{p}\right)e^{2\pi i r/p} \cdot \left(\frac{p}{q}\right)\sum_{s=1}^{q-1}\left(\frac{s}{q}\right)e^{2\pi i s/q} \quad \text{(by Lemma 6.3)}$$

$$= \left(\frac{p}{q}\right)\left(\frac{q}{p}\right)S(p)S(q) \quad \text{(by (6.4))}. \qquad \blacksquare$$

There are many ways to evaluate $S(n)$ now known, both analytic and algebraic. We shall present a twentieth century proof discovered by Schur that is based on the finite Fourier transform.

For the rest of this section we fix once for all a positive odd integer n.

Let $C(\mathbb{Z}/n)$ be the set of all complex valued functions $f\colon \mathbb{Z}/n \to \mathbb{C}$. Thus $C(\mathbb{Z}/n)$ is an n-dimensional complex vector space.

Let $\omega = e^{2\pi i/n}$, so that $S(n) = \sum_{k=0}^{n-1}\omega^{k^2}$.

Definition 6.5. The *Fourier transform* F on $\mathbb{Z}/n$ is the linear map F: $C(\mathbb{Z}/n) \to C(\mathbb{Z}/n)$ defined by

$$Ff(a) = \left(1/\sqrt{n}\right) \sum_{k=0}^{n-1} f(k)\omega^{ka} \quad \text{for } f \in C(\mathbb{Z}/n),\ a \in \mathbb{Z}/n.$$

The evaluation Theorem 6.2 of $S(n)$ is an immediate consequence of Schur's theorem, which follows.

Theorem 6.6. Schur. i. Trace$(F) = (1/\sqrt{n})S(n)$.

ii. The eigenvalues of F are $1, i, -1, -i$. The multiplicities of the eigenvalues (in the preceding order) are

$$\begin{array}{ll} x+1, x, x, x & \text{if } n = 4x+1 \equiv 1 \pmod 4 \\ x, x, x, x-1 & \text{if } n = 4x-1 \equiv 3 \pmod 4 \end{array}.$$

Proof. i. We compute the trace of F by representing F as a matrix. To this end, for $b \in \mathbb{Z}/n$ let $\delta_b \in C(\mathbb{Z}/n)$ be the characteristic function on $\{b\}$; that is, let $\delta_b(a) = 1$ if $a = b$, $\delta_b(a) = 0$ if $a \neq b$. The set $B = \{\delta_b | b \in \mathbb{Z}/n\}$ is a basis for $C(\mathbb{Z}/n)$.

Since

$$F\delta_b(a) = \frac{1}{\sqrt{n}} \sum_{k=0}^{n-1} \delta_b(k)\omega^{ka} = \frac{1}{\sqrt{n}}\omega^{ab},$$

we have

$$F\delta_b = \sum_{a \in \mathbb{Z}/n} \frac{1}{\sqrt{n}} \omega^{ab}\delta_a.$$

Therefore, the matrix of F with respect to the basis B is $1/\sqrt{n}\,(\omega^{ab})_{0 \le a,\, b \le n-1}$. Thus

$$\operatorname{tr}(F) = \sum_{k=0}^{n-1} \frac{1}{\sqrt{n}} \omega^{k^2} = \frac{1}{\sqrt{n}} S(n).$$

ii. Begin with the computation $F^2 f(a) = f(-a)$ for all $f \in C(\mathbb{Z}/n)$, $a \in \mathbb{Z}/n$. Indeed,

$$\begin{aligned} F^2 f(a) &= \frac{1}{\sqrt{n}} \sum_{k \in \mathbb{Z}/n} \left(\frac{1}{\sqrt{n}} \sum_{j \in \mathbb{Z}/n} f(j)\omega^{jk} \right) \omega^{ka} \\ &= \frac{1}{n} \sum_{j \in \mathbb{Z}/n} f(j) \left(\sum_{k \in \mathbb{Z}/n} \omega^{k(j+a)} \right). \end{aligned}$$

The inner sum is 0 by Lemma 6.3 unless $j \equiv -a \pmod{n}$, in which case it equals n. The result follows.

We make two deductions:

First, since $F^4 = (F^2)^2$ is the identity map on $C(\mathbb{Z}/n)$, the eigenvalues of F must all be fourth roots of 1. Let the multiplicities of the eigenvalues $1, i, -1, -i$ of F be m_0, m_1, m_2, m_3. (m_t is the multiplicity of i^t.) Because F has n eigenvalues,

$$m_0 + m_1 + m_2 + m_3 = n. \qquad (*)$$

Second, since $F^2\delta_b = \delta_{-b}$ and $\delta_b \neq \delta_{-b}$ for $b \neq 0 \in \mathbb{Z}/n$ (n is odd), we have that $\mathrm{tr}(F^2) = 1$, which can be written

$$m_0 - m_1 + m_2 - m_3 = 1. \qquad (**)$$

Combining $(*)$ and $(**)$ we find

$$2(m_0 + m_2) = n + 1, \qquad 2(m_1 + m_3) = n - 1. \qquad (***)$$

Next we show that $|S(n)| = \sqrt{n}$. Indeed,

$$\bar{S}S = \sum_{k=0}^{n-1} \omega^{-k^2}\left(\sum_{j=0}^{n-1} \omega^{(j+k)^2}\right) = \sum_{j=0}^{n-1} \omega^{j^2}\left(\sum_{k=0}^{n-1} \omega^{2jk}\right).$$

The last inner sum equals n if $n|2j$, which can only happen when $j = 0$ (because n is odd). Otherwise, by Lemma 6.3, the inner sum equals 0. Thus $|S(n)|^2 = \bar{S}S = n$. From Theorem 6.6i we deduce the relation $(m_0 - m_2)^2 + (m_1 - m_3)^2 = 1$.

There are two cases to be considered. If $m_1 = m_3 = x$ and $m_0 - m_2 = \pm 1$, then $n - 1 = 4x$ by $(***)$ and so $n \equiv 1 \pmod 4$. If $m_0 = m_2 = x$ and $m_1 - m_3 = \pm 1$, then similarly $n + 1 = 4x$ and thus $n \equiv 3 \pmod 4$. Summarizing what we have proved so far, we can write

$$\begin{array}{lll} m_0 - m_2 = \pm 1, & m_1 = m_3 = x & \text{if } n = 4x + 1 \equiv 1 \pmod 4 \\ m_0 = m_2 = x, & m_1 - m_3 = \pm 1 & \text{if } n = 4x - 1 \equiv 3 \pmod 4 \end{array}.$$

It remains but to settle a question of sign. We must show that in the two cases we have $m_0 - m_2 = 1$, $m_1 - m_3 = 1$. This will be done by computing the determinant of F in two ways.

To begin with, the determinant may be calculated as the product of the eigenvalues of F. Thus

$$\det(F) = i^{m_1-m_3-2m_2} = i^{(m_1-m_3)+(m_0-m_2)-(1+n)/2} \quad (\text{by } (***))$$

$$= \begin{cases} (m_0 - m_2)i^{(1-n)/2} & \text{if } n \equiv 1 \pmod 4 \\ (m_1 - m_3)i^{(1-n)/2} & \text{if } n \equiv 3 \pmod 4 \end{cases}. \tag{6.7}$$

On the other hand, we showed in the proof of Theorem 6.6i that the matrix of F with respect to a suitable basis is essentially a Vandermonde matrix, so its determinant may be computed directly:

$$\det(F) = \det\left(\frac{1}{\sqrt{n}}\omega^{ab}\right)_{0\le a,\, b\le n-1} = n^{-n/2} \prod_{0\le a<b\le n-1} (\omega^b - \omega^a)$$

$$= n^{-n/2} \prod_{a<b} e^{2\pi i(a+b)/2n}\left(e^{2\pi i(b-a)/2n} - e^{2\pi i(a-b)/2n}\right)$$

$$= n^{-n/2} e^{(\pi i/n)\Sigma_{a<b}(a+b)} \prod_{a<b}\left(2i \sin\frac{(b-a)\pi}{n}\right).$$

This last expression simplifies. Consider the set $X = \{(a, b) | 0 \le a < b \le n-1)\}$. For $j = 0, 1, \dots, n-1$ there are exactly $n-1$ elements $(a, b) \in X$ with $a = j$ or $b = j$. Thus

$$\sum_{0\le a<b\le n-1} (a+b) = \sum_{j=0}^{n-1} (n-1)j = \left(\frac{n-1}{2}\right)^2 2n \equiv 0 \pmod{2n}.$$

Hence

$$\det(F) = n^{-n/2} \prod_{a<b}\left(2i \sin\frac{(b-a)\pi}{n}\right) = Ki^{n(n-1)/2} \tag{6.8}$$

$$= Ki^{(1-n)/2},$$

where K is a *positive* real number because every sine in the product is positive.

Finally, comparison of the two expressions (6.7) and (6.8) for the determinant of F proves that $m_0 - m_2 = +1$ if $n \equiv 1 \pmod 4$ and that $m_1 - m_3 = +1$ if $n \equiv 3 \pmod 4$. The proof of Schur's theorem is complete. ■

Exercises

1. Let n be a positive odd integer.

i. Show that Ff is an even function of $\mathbb{Z}/n$ if f is even and that Ff is odd if f is odd. Determine the dimensions of the spaces V^+ of even functions and V^- of odd functions in $C(\mathbb{Z}/n)$.

ii. Deduce relation $(***)$ by consideration of the eigenvalues of F when restricted to each of the spaces V^+ and V^-.

2. Let

$$T = \begin{pmatrix} 1 & 1 & 1 & & 1 \\ 1 & x & x^2 & \cdots & x^{n-1} \\ \vdots & & & & \\ 1 & x^{n-1} & x^{(n-1)2} & & x^{(n-1)(n-1)} \end{pmatrix} = (x^{ab})_{0 \le a,\, b \le n-1},$$

where n is a positive integer. Prove that $\det T = \prod_{0 \le a < b \le n-1}(x^b - x^a)$.

3. Show that the linear transformation F is diagonalizable for all odd integers $n > 0$.

4. Let G be a finite abelian group. A *character* of G is a homomorphism χ: $G \to \mathbb{C}^\times$. Let $C(G)$ be the space of all complex valued functions on G and let $\hat{G}$ be the set of characters of G.

i. Show that $\hat{G}$ is a finite abelian group under ordinary multiplication of functions. Note that $\chi^{-1} = \bar{\chi}$ for every $\chi \in \hat{G}$, where $\bar{\chi}$ denotes the function complex conjugate to χ.

ii. Define an inner product $\langle \cdot , \cdot \rangle$ on $C(G)$ by the formula $\langle f, g \rangle = \sum_{x \in G} f(x)\bar{g}(x)$. Show that for $\chi, \chi' \in \hat{G}$, we have

$$\langle \chi, \chi' \rangle = \begin{cases} 0 & \text{if } \chi \neq \chi' \\ |G| & \text{if } \chi = \chi' \end{cases}.$$

Conclude that the set $\hat{G}$ is linearly independent.

iii. Suppose that G is cyclic of order n. Show that there are exactly n characters of G. Conclude that $\hat{G}$ is a vector space basis of $C(G)$.

iv. Prove that $\hat{H} \times \hat{K} \simeq (H \times G)\hat{}$ for every pair of finite abelian groups H and K. Using the structure theorem for finite abelian groups, prove that $\hat{G}$ is a basis of $C(G)$ for every finite abelian group G.

5. Let p be an odd prime number, let $G = U_p \subset \mathbb{Z}/p$, and let $\omega = e^{2\pi i/p}$. We will identify $\hat{G}$ with a subset of $C(\mathbb{Z}/p)$ by "extension by zero," i.e., by defining $\chi(0) = 0$ for all $\chi \in \hat{G}$.

i. Let $\chi \neq 1 \in \hat{G}$ and define $g(\chi) = \sum_{a=1}^{p-1}\chi(a)\omega^a$. Show that $F\chi = (g(\chi)/\sqrt{p})\bar{\chi}$, that $F(F\chi) = \chi(-1)\chi$, and that $g(\chi)g(\bar{\chi}) = \chi(-1)p$.

ii. Let $\chi_1 \in \hat{G}$ be the identity character and let $\chi_2 \in \hat{G}$ be the Legendre symbol, $\chi_2(a) = (a/p)$ for $a \in G$. Show that χ_1 and χ_2 are the only real-valued characters of G.

iii. Order the basis $\hat{G} \cup \{\delta_0\}$ of $C(\mathbb{Z}/p)$ as follows: first δ_0, then χ_1, then χ_2, then the rest of the elements $\chi, \bar{\chi}, \chi', \bar{\chi}', \ldots$ of $\hat{G}$ arranged in pairs of complex conjugate characters. Prove that the matrix of F with respect to the preceding ordered basis is the diagonal block matrix

$$F = \frac{1}{\sqrt{p}}\operatorname{diag}\left(\begin{pmatrix} 1 & p-1 \\ 1 & -1 \end{pmatrix}, (g(\chi_2)), \begin{pmatrix} 0 & g(\bar{\chi}) \\ g(\chi) & 0 \end{pmatrix}, \begin{pmatrix} 0 & g(\bar{\chi}') \\ g(\chi') & 0 \end{pmatrix}, \ldots\right).$$

iv. Compute $\det(F)$ from the matrix expression of iii. Compare with (6.8) to evaluate $g(\chi_2) = S(p)$.

v. Determine the eigenvectors of F in $C(\mathbb{Z}/p)$.

7. The Ring $\mathbb{Z}[e^{2\pi i/n}]$

The computations of Gauss's sixth proof of quadratic reciprocity take place within a natural generalization of the ring $\mathbb{Z}[i]$ of Gaussian integers.

Definition 7.1. For every positive integer n define the *cyclotomic ring* $\mathbb{Z}[e^{2\pi i/n}] = \{\sum_{k=0}^{n-1} m_k e^{2\pi i k/n} | m_k \in \mathbb{Z}\} \subset \mathbb{C}$.

The cyclotomic ring $\mathbb{Z}[e^{2\pi i/n}]$ was originally investigated in connection with Fermat's Diophantine equation $X^n + Y^n = Z^n$, still a subject of active research. Of greater historical significance perhaps is the major role played by the cyclotomic rings in the development of both algebraic number theory and modern ring theory. These rings were at first vexing because in most cases factorization of cyclotomic numbers into products of prime elements in the rings is not unique. They belong to the class of rings now called Dedekind domains. We shall not go into this interesting story here, since our present interest is just a proof of the Law of Quadratic Reciprocity. For that we need just one simple property of the cyclotomic rings.

Proposition 7.2. $\mathbb{Z}[e^{2\pi i/n}] \cap \mathbb{Q} = \mathbb{Z}$ for all $n \geq 1$.

Proof. We will prove first that every element of $\mathbb{Z}[e^{2\pi i/n}]$ is the zero of a monic polynomial with coefficients in $\mathbb{Z}$. Then we show that every rational number with the same property is actually an integer. Let $\omega = e^{2\pi i/n}$.

Let $z \in \mathbb{Z}[\omega]$. Since $\mathbb{Z}[\omega]$ is a ring, $z\omega^i \in \mathbb{Z}[\omega]$ for every integer i. Thus there exist integers a_{ij} such that $z\omega^i = \sum_{j=0}^{n-1} a_{ij}\omega^j$, $0 \le i \le n-1$. Setting $A = (a_{ij})_{0 \le i, j \le n-1}$, we can write these equations as a single matrix equation: $Av = zv$, where $v = {}^t(\omega^0, \omega^1, \dots, \omega^{n-1})$. In other words, z is an eigenvalue of the $n \times n$ matrix A. Therefore, z is a root of the characteristic polynomial of A, which is monic and has coefficients in $\mathbb{Z}$ because A has integer entries.

Let $r = u/v \in \mathbb{Q}$, where u and v are relatively prime integers, $v \ne 0$. Suppose that r is a zero of a polynomial $p(X) = X^n + \sum_{i=1}^n a_i X^{n-i} \in \mathbb{Z}[X]$. Then $v^n p(r) = u^n + \sum_{i=1}^n a_i u^{n-i} v^i = 0$, which shows that $v | u^n$. Since $\mathrm{GCD}(u, v) = 1$, it follows that $v|u$. Thus $r = u/v \in \mathbb{Z}$. ∎

***Definition* 7.3.** For odd prime numbers p, set $g_p = \sum_{a=1}^{p-1}(a/p)e^{2\pi i a/p}$.

The sum g_p is called a Gauss sum. In view of (6.4) we have $g_p = S(p)$, where the number $S(p) \in \mathbb{Z}[e^{2\pi i/p}]$ was evaluated in Theorem 6.2. But the arguments to follow will in no way depend upon that previous evaluation.

Lemma 7.4. Let p and q be distinct odd primes.

i. $(g_p)^2 = p^*$.

ii. $(g_p)^{q-1} \equiv (q/p) \pmod q$.

Proof. Let $\omega = e^{2\pi i/p}$.

i. In view of (6.4), this is an immediate consequence of Theorem 6.2. But a simple proof by direct computation can be given as follows:

$$(g_p)^2 = \sum_{a=1}^{p-1}\left(\frac{a}{p}\right)\omega^a \sum_{b=1}^{p-1}\left(\frac{b}{p}\right)\omega^b = \sum_{a=1}^{p-1}\left(\left(\frac{a}{p}\right)\omega^a \sum_{b=1}^{p-1}\left(\frac{ab}{p}\right)\omega^{ab}\right)$$

$$= \sum_{b=1}^{p-1}\left(\frac{b}{p}\right)\sum_{a=1}^{p-1}\omega^{a(b+1)}.$$

By Lemma 6.3. we have that

$$\sum_{a=1}^{p-1}\omega^{a(b+1)} = \begin{cases} p-1 & \text{if } b \equiv -1 \pmod p \\ -1 & \text{otherwise} \end{cases}.$$

Thus

$$(g_p)^2 = \left(\frac{p-1}{p}\right)(p-1) - \left(\sum_{b=1}^{p-1}\left(\frac{b}{p}\right) - \left(\frac{p-1}{p}\right)\right) = \left(\frac{-1}{p}\right)p = p^*,$$

where we have used Proposition 3.2i, the fact that there are exactly as many quadratic residues as quadratic nonresidues mod p.

ii. The assertion to be proved has meaning because i implies that $(g_p)^{q-1} \in \mathbb{Z}$.

We will write $A \equiv B \pmod{q\mathbb{Z}[\omega]}$ for complex numbers A and B to indicate that $A - B \in q\mathbb{Z}[\omega]$. Compute as follows:

$$\begin{aligned}
(g_p)^q &= \left(\sum_{a=1}^{p-1} \left(\frac{a}{p}\right)\omega^a\right)^q \equiv \sum_{a=1}^{p-1}\left(\frac{a}{p}\right)\omega^{aq} \quad (\mathrm{mod}\ q\mathbb{Z}[\omega]) \\
&= \sum_{a=1}^{p-1}\left(\frac{ax}{p}\right)\omega^a \quad \text{where } x \in \mathbb{Z} \text{ satisfies } xq \equiv 1 \pmod{p} \\
&= \left(\frac{q}{p}\right) g_p.
\end{aligned}$$

The congruence mod $q\mathbb{Z}[\omega]$ holds because most of the multinomial coefficients in a q-power expansion are divisible by the prime q.

Multiplying by $g_p \in \mathbb{Z}[\omega]$ and using i, we find that $p^* g_p^{q-1} \equiv q^*(q/p)$ $(\mathrm{mod}\ q\mathbb{Z}[\omega])$. Since both sides of this congruence lie in $\mathbb{Z}$, Proposition 7.2 proves that $p^* g_p^{q-1} \equiv p^*(q/p) \pmod{q\mathbb{Z}}$. We get what we want by cancelling p^* from this congruence, which is possible because p and q are relatively prime (Lemma 3.6 of Chapter 1). ■

We can now prove quadratic reciprocity in one line.

Proof of Proposition 1.2. We will prove Proposition 3.4iii. By Proposition 3.2ii and Lemma 7.4 we have

$$\left(\frac{p^*}{q}\right) \equiv (p^*)^{(q-1)/2} = (g_p)^{q-1} \equiv \left(\frac{q}{p}\right) \pmod{q}.$$ ■

Exercise

1. Let p be an odd prime, let $\omega = e^{2\pi i/8}$, and let $g = \omega + \omega^{-1}$.
   i. Show that $g^2 = 2$.
   ii. Show that $g^p \equiv \omega^p + \omega^{-p} \pmod{p\mathbb{Z}[\omega]}$.
   iii. Show that $\omega^p + \omega^{-p} = \pm g$ and determine how the sign depends on p.
   iv. Prove that

$$\left(\frac{2}{p}\right) = \begin{cases} 1 & \text{if } p \equiv \pm 1 \pmod{8} \\ -1 & \text{if } p \equiv \pm 3 \pmod{8} \end{cases}.$$

8. The Jacobi Symbol

For computational purposes it is convenient to extend the domain of the Legendre symbol.

Definition 8.1. The *Jacobi symbol* (a/m) is defined for all $a, m \in \mathbb{Z}$ such that m is odd and positive and $\text{GCD}(a, m) = 1$ by the equation $(a/m) = \prod_i (a/p_i)$, where $m = \prod_i p_i$, the p_i are prime, and (a/p_i) denotes the Legendre symbol.

The important properties of the Jacobi symbol are all easily deducible from its definition and the properties of the Legendre symbol. Note particularly multiplicativity in both arguments $(ab/m) = (a/m)(b/m)$ and $(a/mn) = (a/m)(b/n)$. Observe also that $(a/m) = (b/m)$ if $a \equiv b \pmod m$. We give a more formal statement of the reciprocity law, which is the same as that of the Legendre symbol.

Proposition 8.2. Jacobi Reciprocity and Supplements. Let m and n be odd positive integers such that $\text{GCD}(m, n) = 1$.

i. $$\left(\frac{-1}{m}\right) = (-1)^{(m-1)/2} = \begin{cases} 1 & \text{if } m \equiv 1 \pmod 4 \\ -1 & \text{if } m \equiv 3 \pmod 4 \end{cases}.$$

ii. $$\left(\frac{2}{m}\right) = (-1)^{(m^2-1)/8} = \begin{cases} 1 & \text{if } m \equiv \pm 1 \pmod 8 \\ -1 & \text{if } m \equiv \pm 3 \pmod 8 \end{cases}.$$

iii. $$\left(\frac{n}{m}\right) = (-1)^{(m-1)/2 \cdot (n-1)/2}\left(\frac{m}{n}\right)$$
$$= \begin{cases} \left(\dfrac{m}{n}\right) & \text{if } m \equiv 1 \text{ or } n \equiv 1 \pmod 4 \\ -\left(\dfrac{m}{n}\right) & \text{if } m \equiv n \equiv 3 \pmod 4 \end{cases}.$$

Equivalently, $(m^*/n) = (n/m)$.

Proof. i. Let $m = \prod_i p_i$, where the p_i are odd prime numbers. Then $(-1/m) = \prod_i (-1/p_i) = \prod_i (-1)^{(p_i-1)/2} = (-1)^{(m-1)/2}$, where we have used Proposition 3.3i and Lemma 3.7iv.

iii. The equivalence of the two assertions of iii is a consequence of i. We prove that $(m^*/n) = (n/m)$.

Let $m = \prod_i p_i$ and let $n = \prod_j q_j$, where the p_i and q_j are odd prime numbers. Using multiplicativity of the Legendre symbol, the Law of Quadratic

Reciprocity 3.4iii, and Lemma 3.7ii, we compute $(m^*/n) = \Pi_i \Pi_j (p_i^*/q_j) = \Pi_i \Pi_j (q_j/p_i) = (n/m)$.

ii. We have $((m+2)/m) = (m/(m+2))$ by iii, since $m \not\equiv m+2 \pmod 4$. Thus $(2/m) = ((m+2)/m) = (m/(m+2)) = (-2/(m+2)) = (-1/(m+2))(2/(m+2))$. In particular, $(2/(8k+1)) = -(2/(8k+3)) = -(2/(8k+5)) = (2/(8k+7)) = (2/(8(k+1)+1))$ for all integers $k \geq 0$. An induction beginning with the case $(2/1) = 1$ shows that $(2/(8k+1)) = 1$ for all $k \geq 0$. Statement ii follows immediately. ∎

It is worth noting that the preceding proof of Proposition 8.2ii did not rely upon a prior evaluation of the Legendre symbols $(2/p)$. It therefore constitutes an independent proof of Proposition 3.3ii.

The properties of the Jacobi symbol that we have derived can be combined into a fast recursive algorithm to compute it. One straightforward procedure to compute (m/n) goes as follows.

1. $(0/1) = 1$.
2. If $m < 0$, then $(m/n) = (-1/n)(|m|/n)$.
3. If $m \geq n$, then $(m/n) = (r/n)$, where $m \equiv r \pmod n$ and $0 \leq r < n$.
4. If $2|m$, then $(m/n) = (2/n) \cdot ((m/2)/n)$.
5. $(m/n) = (n^*/m)$.

Much of the importance of the Jacobi symbol derives from the fact that it can be computed efficiently. The Legendre symbol is a special case, so it can therefore be computed quickly too. The answer to the question of the existence of solutions of the congruence $X^2 \equiv a \pmod p$ for fixed a and prime number p can therefore be found quickly by computing (a/p). A trial and error search for solutions through $X = 1, 2, \ldots, (p-1)/2$ would almost always be totally impractical for very large p.

The rest of this section is a digression on a practical method to discover large primes that was found by Solovay and Strassen.

Let m be an odd prime number. Then every $x = 1, 2, \ldots, m-1$ satisfies the two conditions (by Proposition 3.2ii)

$$\mathrm{GCD}(x, m) = 1, \tag{8.3a}$$

$$x^{(m-1)/2} \equiv \left(\frac{x}{m}\right) \pmod m. \tag{8.3b}$$

The Solovay–Strassen test is based on a strong converse statement.

Definition 8.4. Let $m > 1$ be an odd integer. We say that $x \in \{1, 2, \ldots, m - 1\}$ is a *witness that m is composite* iff either of the conditions (8.3a) or (8.3b) fails for x.

To prove that an odd integer is composite, it clearly suffices to find a witness. But how easy is that to do?

Proposition 8.5. Let $m > 1$ be a composite odd integer. Then more than half the integers $1, 2, \ldots, m - 1$ are witnesses that m is composite.

Proof. Let $m > 1$ be a composite odd integer. We prove first that there is a witness that m is composite that is relatively prime to m.

Suppose first that there is a prime number p such that $p^2 | m$. Then there exists an element x of U_m that has order p. One can take for instance $x = 1 + m/p \in U_m$, for it follows easily from the binomial theorem that $(1 + m/p)^p \equiv 1 \pmod{m}$. We can compute $(x/m)^2 = (\pm 1)^2 = 1$. But $(x^{(m-1)/2})^2 = x^{m-1} \not\equiv 1 \pmod{m}$ because p, the order of x in U_m, does not divide $m - 1$. Thus x is a witness.

On the other hand, suppose that $m = p_1 p_2 \cdots p_r$, where the p_i are distinct odd primes and $r \geq 2$. Let $a \in \mathbb{Z}$ be such that $p_1 \nmid a$ and $(a/p_1) = -1$. By the Chinese Remainder Theorem 2.2 choose $x \in \mathbb{Z}$ such that $x \equiv a \pmod{p_1}$ and $x \equiv 1 \pmod{p_i}$ for $i \geq 2$. Then $(x/m) = \prod_i (x/p_i) = -1$, but $x^{(m-1)/2} \equiv 1 \pmod{p_2}$. Thus x is a witness.

Let $\phi\colon U_m \mapsto U_m$ be the homomorphism $\phi(x) = x^{(m-1)/2} \cdot (x/m)$. The kernel of ϕ is the set of elements of $\{1, 2, \ldots, m - 1\}$ that are *not* witnesses that m is composite. We have just proved that ϕ is nontrivial, so the kernel of ϕ is a proper subgroup of U_m. Its index in U_m must be at least 2. Hence $|\ker \phi| \leq |U_m|/2 < (m - 1)/2$. Therefore, the number of witnesses that m is composite must be greater than $(m - 1)/2$. ∎

The Solovay–Strassen test that an integer is prime is simply a random search for witnesses that it is composite. If no witness is found, the test asserts, possibly wrongly, that the integer is prime. It can be analyzed as follows.

Let $m > 1$ be a composite odd integer. Proposition 8.5 shows that a random search among the integers $1, 2, \ldots, m - 1$ will very probably turn up a witness that m is composite, and hence a proof of the fact, fairly quickly. Therefore failure to discover a witness that an integer is composite after such a random search can be overwhelming circumstantial evidence that the integer is prime. It's as if you were given a coin that you knew to be either a normal penny or a two-headed freak. Suppose you toss the coin 100 times, getting heads each time. What would you say?

To *prove* that a suspected prime number is actually prime requires quite different techniques.

Exercises

1. Find a pair of positive relatively prime integers a and m such that $(a/m) = 1$ but $X^2 \equiv a \pmod{m}$ has no solution.

2. Evaluate $(3989/144169)$.

3. Prove Proposition 8.2ii directly from Proposition 3.3ii and Definition 8.1.

4. Let p be an odd prime and let $a \in \mathbb{Z}$ with $p \nmid a$.

i. Prove that if $p \equiv 3 \pmod 4$, then the congruence $X^4 \equiv a \pmod{p}$ has a solution if and only if $(a/p) = 1$.

ii. Prove that if $p \equiv 1 \pmod 4$, then the equation $X^4 \equiv a \pmod{p}$ has a solution if and only if $a^{(p-1)/4} \equiv 1 \pmod{p}$.

5. Find all primes p such that following congruences have solutions.

i. $X^4 \equiv 4 \pmod{p}$.

ii. $X^4 \equiv -4 \pmod{p}$.

6. Let $p \equiv 1 \pmod 4$ be prime. Write $p = a^2 + b^2$, where a, b are positive integers and a is odd. Let $f \in \mathbb{Z}$ satisfy the congruence $b \equiv af \pmod{p}$. Prove the following:

i. $((a+b)/p) \equiv (2ab)^{(p-1)/4} \pmod{p}$.

ii. $((a+b)/p) = (2/(a+b))$. (*Hint:* $2p \equiv (a-b)^2 \pmod{a+b}$.)

iii. $f^2 \equiv -1 \pmod{p}$.

iv. $2^{(p-1)/4} \equiv f^{ab/2}$.

v. $X^4 \equiv 2 \pmod{p}$ has a solution if and only if $b \equiv 0 \pmod 8$ if and only if $X^2 + 64Y^2 = p$ has an integral solution.

7. Let $a, b \in \mathbb{Z}$, where $\mathrm{GCD}(a, b) = 1$. Suppose that $a \equiv b \equiv 3 \pmod 4$ and that there exist $\alpha, \beta \in \mathbb{Z}$ such that $\beta^2 \equiv b \pmod{a}$ and $\alpha^2 \equiv a \pmod{b}$. Prove that $a < 0$ and $b < 0$. (This exercise shows that the sign of an integer and its congruence properties are not independent.)

9. The Kronecker Symbol

Definition 9.1. An integer Δ is said to be a *discriminant* iff $\Delta \equiv 0$ or $1 \pmod 4$.

Definition 9.2. The *Kronecker symbol* χ_Δ is defined for nonsquare discriminants Δ to be the function on $\{m \in \mathbb{Z} | \mathrm{GCD}(m, \Delta) = 1\}$ given by

$$\chi_\Delta(m) = \begin{cases} (m/|\Delta|) & \text{if } \Delta \text{ is odd} \\ (-1)^{(d-1)/2 \cdot (m-1)/2}(2^a/|m|)(m/|d|) & \\ & \text{if } \Delta \text{ is even, } \Delta = 2^a d,\ d \text{ odd} \end{cases}$$

Notice an old friend, $\chi_{-4}(m) = (-1)^{(m-1)/2}$ for odd integers m.

The Kronecker symbol is just as easy to evaluate as the Jacobi symbol.

Proposition 9.3. Let Δ be a nonsquare discriminant.

i. Let $m, n \in \mathbb{Z}$ be relatively prime to Δ. Then $\chi_\Delta(mn) = \chi_\Delta(m)\chi_\Delta(n)$. If $m \equiv n \pmod{\Delta}$, then $\chi_\Delta(m) = \chi_\Delta(n)$.
ii. $\chi_\Delta(m) = (\Delta/m)$ for all positive odd integers m relatively prime to Δ.
iii. $\chi_\Delta(-1) = \operatorname{sign} \Delta$.
iv. $\chi_\Delta(2) = \begin{cases} 1 & \text{if } \Delta \equiv 1 \pmod 8 \\ -1 & \text{if } \Delta \equiv 5 \pmod 8 \end{cases}.$
v. There exists $m \in \mathbb{Z}$ relatively prime to Δ such that $\chi_\Delta(m) = -1$.
vi. The function $\chi_\Delta\colon U_\Delta \to \{\pm 1\}$ is a surjective group homomorphism.

Proof. i. The multiplicativity of χ_Δ follows from Lemma 3.7iv and the multiplicativity of the Jacobi symbol in both arguments.

The second assertion requires comment only when Δ is even. Suppose that $\Delta = 2^a d$ is even, with d odd and $a \geq 2$, and that $m \equiv n \pmod{\Delta}$. Then $m \equiv n \pmod 4$, which shows that $(-1)^{(m-1)/2} = (-1)^{(n-1)/2}$. If $a = 2$, then $(2^a/|m|) = (2^a/|n|) = 1$. If $a \geq 3$, then $m \equiv n \pmod 8$, whence $|m| \equiv \pm|n| \pmod 8$, and thus by Proposition 8.2ii we have $(2^a/|m|) = (2^a/|n|)$. Finally, the congruence $m \equiv n \pmod d$ implies that $(m/|d|) = (n/|d|)$. It follows that $\chi_\Delta(m) = \chi_\Delta(n)$.

ii. Use the reciprocity law.

For odd Δ,

$$(m/|\Delta|) = (|\Delta|^*/m) = (\Delta/m),$$

since $\Delta = |\Delta|^*$.

For even $\Delta = 2^a d$,

$$(m/|d|) = (|d|^*/m) = (-1)^{(d-1)/2 \cdot (m-1)/2}(d/m)$$

and $(2/|m|)^a = (2/m)^a$. Use multiplicativity of the Jacobi symbol.

iii. For odd Δ we have

$$(-1/|\Delta|) = (-1)^{(|\Delta|-1)/2} = \operatorname{sign}(|\Delta| \cdot |\Delta|^*) = \operatorname{sign} \Delta,$$

since $|\Delta| > 0$ and $|\Delta|^* = \Delta$.

For even $\Delta = 2^a d$,

$$\chi_\Delta(-1) = (-1)^{(d-1)/2}(-1/|d|) = \operatorname{sign}(d \cdot d^* \cdot |d| \cdot |d|^*)$$
$$= \operatorname{sign}(d) = \operatorname{sign} \Delta,$$

since $d^*|d|^* = d^2 > 0$.

iv. If $\Delta \equiv 1 \pmod 8$, then $|\Delta| \equiv \pm 1 \pmod 8$. If $\Delta \equiv 5 \pmod 8$, then $|\Delta| \equiv \pm 3 \pmod 8$. Apply Proposition 8.2ii.

v. If $\Delta < 0$, then $\chi_\Delta(-1) = -1$, so suppose that $\Delta > 0$.

We are going to use the Chinese Remainder Theorem 2.2 to produce an integer m such that $\chi_\Delta(m) = -1$.

Write $\Delta = 2^a d$, where d is odd. Since Δ is not a square, either a is odd or d is not a square. For positive odd integers m we have $\chi_\Delta(m) = (2^a d/m) = (2/m)^a (m^*/d)$.

If a is odd, choose an integer $m > 0$ such that $m \equiv 5 \pmod 8$ and $m \equiv 1 \pmod d$. Then $m = m^*$ and so $\chi_\Delta(m) = (-1)^a(1/d) = -1$.

If a is even, then d is not a square. Thus there is a prime number p such that $d = p^r f$ where r is odd, $f \in \mathbb{Z}$, and $p \nmid f$. Let $b \in \mathbb{Z}$ be such that $p \nmid b$ and $(b/p) = -1$. Choose an integer $m > 0$ such that $m \equiv b \pmod p$ and $m \equiv 1 \pmod{4f}$. Then $m = m^*$. We have $\chi_\Delta(m) = (m/p)^r(m/f) = (b/p)^r(1/f) = (-1)^r = -1$.

vi. By i and v. ■

With the introduction of the Kronecker symbol we can at last give a satisfactory answer to the question: Given an integer a, for which prime numbers p does there exist a solution to the congruence equation $X^2 \equiv a \pmod p$?

If a is a square, the answer is of course that solutions exist for all primes p. If a is not a discriminant, then $4a$ is a discriminant and the two congruences $X^2 \equiv a$ and $X^2 \equiv 4a \pmod p$ have solutions for exactly the same sets of primes p. So the case of real interest is that of a nonsquare discriminant a.

Suppose that $a \in \mathbb{Z}$ is a nonsquare discriminant and that p is an odd prime number not dividing a. Then $\chi_a(p) = (a/p)$. We can assert that $X^2 \equiv a \pmod{p}$ has a solution if and only if $\chi_a(p) = 1$. The remarkable features of this assertion are the properties Proposition 9.3i of the function χ_a, namely that χ_a is periodic mod a and multiplicative. We summarize this in our next statement, which includes Theorem 1.3.

Theorem 9.4. Let $\Delta \in \mathbb{Z}$ be a nonsquare discriminant.

Then χ_Δ is the unique homomorphism $f\colon U_\Delta \to \{\pm 1\}$ such that the following two conditions are equivalent for all odd primes p that do not divide Δ.

1. $X^2 \equiv \Delta \pmod{p}$ has a solution.
2. $\bar{p} \in \ker f$.

Therefore, the primes p for which $X^2 \equiv \Delta \pmod{p}$ has a solution are precisely those odd primes that lie in a fixed subgroup of U_Δ of index 2 together with the prime divisors of 2Δ.

Proof. The deep question of existence of the homomorphism f is settled by the construction of the Kronecker symbol. Gauss's Law of Quadratic Reciprocity was the key ingredient in this development.

Uniqueness is much easier. The equivalence of conditions 1 and 2 specifies the values $f(\bar{p}) = (\Delta/p)$ for odd primes p not dividing Δ. But this determines f completely because these elements $\bar{p}$ are easily seen to form a set of generators for the group U_Δ. Let $x \in U_\Delta$. There exists a positive odd integer a such that $x = \bar{a}$. Let $a = \prod_i p_i$, where the p_i are prime. Then $x = \prod_i \bar{p}_i$. ∎

Theorem 9.4 is a straight generalization of Lemma 1.4 of Chapter 2, which analyzed the congruence $X^2 \equiv -1 \pmod{p}$ for odd prime numbers p. That congruence was the starting point for our study of the specific quadratic form $X^2 + Y^2$. With Theorem 9.4 in hand we are in a position to take the first step in the general theory of the representation of prime numbers by integral binary quadratic forms.

Exercises

1. Let Δ be a nonsquare even discriminant. Show that $\chi_\Delta(m) = (\operatorname{sign} \Delta)^{(m-1)/2}(|\Delta|/|m|)$ for all integers m relatively prime to Δ.

2. Determine $\ker \chi_\Delta$ for all nonsquare discriminants Δ such that $|\Delta| \leq 15$.

3. Prove that $U_\Delta \simeq \ker \chi_\Delta \times \{\pm 1\}$ for all negative discriminants Δ.

4. Let Δ be a nonsquare discriminant and let $f \neq 0 \in \mathbb{Z}$. Prove that $\chi_\Delta(m) = \chi_{f^2\Delta}(m)$ for all integers m such that $\text{GCD}(m, f\Delta) = 1$.

5. Let Δ be a nonsquare discriminant.

i. Prove that there is a prime $p \nmid \Delta$ such that $\chi_\Delta(p) = -1$.

ii. Prove that there is an infinite number of primes $p \nmid \Delta$ such that $\chi_\Delta(p) = -1$. (*Suggestion:* Use Exercise 4.)

iii. What does ii say for $\Delta = -4$ and $\Delta = 5$?

6. Let $a \in \mathbb{Z}$. Show that if $X^2 = a$ has no integral solution, then there exists an infinite number of primes p such that $X^2 \equiv a \pmod{p}$ has no solution.

7. Using Dirichlet's Theorem on Primes in Arithmetic Progressions, prove that χ_Δ is the unique *function* $f\colon U_\Delta \to \{\pm 1\}$ such that $X^2 \equiv \Delta \pmod{p}$ has a solution for an odd prime p not dividing Δ if and only if $f(\bar{p}) = 1$.

8. Let q be an odd prime. Show that $\ker \chi_{4q} = \{\pm x^2 | x \in U_{4q}\}$.

10. Binary Quadratic Forms

The study of quadratic congruences was undertaken with binary quadratic forms in mind. We now make the connection between the two subjects.

Proof of Proposition 1.1. $1 \Rightarrow 2$. Suppose that an integral form $F(X, Y) = aX^2 + bXY + cY^2$ of discriminant Δ represents a prime number p, say $F(r, s) = p$, where $r, s \in \mathbb{Z}$. Since p is prime, $\text{GCD}(r, s) = 1$. Thus there exist integers t, u such that

$$g = \begin{pmatrix} r & s \\ t & u \end{pmatrix} \in \mathbf{SL}_2(\mathbb{Z}).$$

Then $gF = pX^2 + b'XY + c'Y^2$, where $b', c' \in \mathbb{Z}$. The equivalent forms F and gF have the same discriminant. Hence $\Delta = b'^2 - 4pc'$, which shows that $b'^2 \equiv \Delta \pmod{p}$.

$2 \Rightarrow 1$. Suppose that $m \in \mathbb{Z}$ solves the congruence $m^2 \equiv \Delta \pmod{p}$. After replacing m by $m + p$ if necessary, we may assume that m and Δ have the same parity (because p is odd). Write $m^2 = \Delta + np$, with $n \in \mathbb{Z}$. Then $np = m^2 - \Delta \equiv 0 \pmod{4}$, because Δ is congruent to 0 or 1 mod 4. Thus $n/4 \in \mathbb{Z}$. The form $pX^2 + mXY + (n/4)Y^2$ is an integral form of discriminant Δ that represents p. ∎

We can now answer the question initially posed in Section 1. The next theorem has been the goal of this chapter.

Theorem 10.1. Let Δ be a nonsquare discriminant and let p be a prime number. The following two conditions are equivalent.

1. There exists an integral binary quadratic form of discriminant Δ that represents p.
2. Either $p|\Delta$ or $p \nmid \Delta$ and $\chi_\Delta(\bar{p}) = 1$, where $\chi_\Delta\colon U_\Delta \to \{\pm 1\}$ is the Kronecker symbol.

Proof. $1 \Rightarrow 2$. If p is odd, the implication is an immediate consequence of Proposition 1.1 and Theorem 9.4.

It remains to examine the case $p = 2$ and $\Delta \equiv 1 \pmod 4$. In view of Proposition 9.3iv, we are to prove that $ax^2 + bxy + cy^2 = 2$ with $a, b, c, x, y \in \mathbb{Z}$ and b odd implies $\Delta = b^2 - 4ac \equiv 1 \pmod 8$. If x is even and y is odd, then c is even. If x is odd and y is even, then a is even. If x and y are both odd, then a and c have opposite parity. In all cases, $\Delta \equiv b^2 \equiv 1 \pmod 8$.

$2 \Rightarrow 1$. If p is odd, the implication is an immediate consequence of Proposition 1.1 and Theorem 9.4.

It remains to examine the case $p = 2$. If $\Delta \equiv 1 \pmod 8$, then $2X^2 + XY + ((1 - \Delta)/8)Y^2$ is an integral form of discriminant Δ that represents 2. If $8|\Delta$, then $2X^2 - (\Delta/8)Y^2$ has discriminant Δ and represents 2. If $\Delta = 4d$ where d is odd, then $2X^2 + 2XY + ((1 - d)/2)Y^2$ has discriminant Δ and represents 2. ∎

Some complements can be given.

We first take up the question of uniqueness of the representing form.

Proposition 10.2. Any two integral forms of the same discriminant that represent the same prime number p must be equivalent forms.

Proof. Let F be an integral binary form that represents a prime number p. In the first part of the proof of Proposition 1.1 we showed that F is properly equivalent to a form F' with leading coefficient equal to p. By choice of $n \in \mathbb{Z}$ we can arrange that $\begin{pmatrix} 1 & 0 \\ n & 1 \end{pmatrix} F' = pX^2 + bXY + cY^2$, where $-p < b \le p$. We will prove that all integral forms $pX^2 + bXY + cY^2$ such that $-p < b \le p$ and with the same discriminant Δ are equivalent. Note that two such forms with the same middle coefficient b are actually equal, because c is determined by the equation $b^2 - 4pc = \Delta$. The parity of b is the same as that of Δ.

Suppose first that p is odd. If $p|\Delta$, then $p|b$, so that $b \in \{0, p\}$. Thus b is uniquely determined by the parity of Δ. If $p \nmid \Delta$, let β be the unique integer with the parity of Δ such that $\beta^2 \equiv \Delta \pmod{p}$ and $0 < \beta < p$. Then $b = \pm\beta$. But the two forms $pX^2 + \beta XY + cY^2$ and $pX^2 - \beta XY + cY^2$ are clearly equivalent.

Now suppose that $p = 2$. Then $-1 \le b \le 2$. If Δ is even, then $b \in \{0, 2\}$ is uniquely determined by the condition $b^2 \equiv \Delta \pmod{8}$. If Δ is odd, then $b = \pm 1$ and the two relevant forms are equivalent. ■

A representation of a prime p by a form $aX^2 + bXY + cY^2$ with coefficients divisible by p might better be thought of as a representation of 1 by the form $(a/p)X^2 + (b/p)XY + (c/p)Y^2$. A definition is suggested.

Definition 10.3. A nonzero integral binary quadratic form $aX^2 + bXY + cY^2$ is *primitive* iff $\text{GCD}(a, b, c) = 1$.

Proposition 10.4. Let p be a prime number that is represented by an integral form F of discriminant Δ. Then F is not primitive if and only if $p^2 | \Delta$ and Δ/p^2 is a discriminant.

Thus the forms of discriminant Δ that represent p are primitive if and only if either (a) $p^2 \nmid \Delta$ or (b) $p = 2$ and $\Delta \equiv 8$ or $12 \pmod{16}$.

Proof. If p divides the three coefficients of an integral binary form F of discriminant Δ, then p^2 must divide Δ and Δ/p^2 is the discriminant of the integral form F/p.

Conversely, suppose that $p^2 | \Delta$ and that Δ/p^2 is a discriminant. There is an integral form G of discriminant Δ/p^2 that represents 1. Then pG has discriminant Δ and represents p. Thus none of the forms of discriminant Δ that represent p are primitive because by Proposition 10.2 they are all equivalent to pG, which is not primitive.

The second assertion is a rewording of the first. ■

Let us summarize the results we have proved.

Let Δ be a nonsquare discriminant. The integral binary quadratic forms of discriminant Δ are partitioned into equivalence classes. The prime numbers are partitioned into two sets: (1) those primes that are represented by no form of discriminant Δ and (2) those primes that are represented by all the forms in one equivalence class of forms of discriminant Δ and by none of the forms in the others. Theorem 10.1, which is really just one form of the Law of Quadratic Reciprocity, gives a satisfactory description of the two sets of primes by congruence conditions.

The preceding theory says nothing about which equivalence classes of forms represent which primes. The problem does not arise for those discriminants Δ with only one equivalence class of primitive forms that represent positive integers (i.e., positive definite if $\Delta < 0$) such as $\Delta = -3, -4, 5, -7, \pm 8, -11, -12, 13$. But in general the situation is very difficult. Gauss made an initial attack on the problem with his class groups and theory of genera. But that is matter for Chapter 5.

Exercises

1. Prove that the form $X^2 + XY + 2Y^2$ represents a prime p if and only if $p = 7$ or $p \equiv 1, 2$, or $4 \pmod 7$.

2. Determine the primes p that are represented by $X^2 + XY + 3Y^2$.

3. Show that every integer is represented by some integral form of every square discriminant.

4. An integral binary quadratic form f is said to represent an integer n *properly* iff there exists $(x, y) \in \mathbb{Z}^2$ such that $\text{GCD}(x, y) = 1$ and $f(x, y) = n$. Let Δ be a discriminant and let $n \neq 0 \in \mathbb{Z}$. Prove that there is a form of discriminant Δ that represents n properly if and only if the congruence $X^2 \equiv \Delta \pmod{4n}$ has a solution.

5. Can equivalence be replaced by proper equivalence in the conclusion of Proposition 10.2?

6. Show that the forms $X^2 + XY + 6Y^2$ and $2X^2 + XY + 3Y^2$ have the same discriminant, are inequivalent, and both represent 4. Thus the hypothesis that p be prime in Proposition 10.2 is essential.

7. **i.** Let $p \neq 2, 5$ be prime. Show that p is represented by a positive definite form of discriminant -20 if and only if $p \equiv 1, 3, 7$, or $9 \pmod{20}$.

ii. Let $p \neq 2, 5$ be prime. Prove that p is represented by $X^2 + 5Y^2$ if and only if $p \equiv 1$ or $9 \pmod{20}$ and that p is represented by $2X^2 + 2XY + 3Y^2$ if and only if $p \equiv 3$ or $7 \pmod{20}$.

iii. Verify $(2x^2 + 2xy + 3y^2)(2z^2 + 2zw + 3w^2) = X^2 + 5Y^2$, where $X = xw + yz - 2yw + 2xz$ and $Y = xw + yz + yw$.

iv. Let $n = 5^b(p_1^{c_1} \cdots p_r^{c_r})(q_1^{d_1} \cdots q_s^{d_s})(u_1^{e_1} \cdots u_t^{e_t})$, where the ps are primes $\equiv 1$ or $9 \pmod{20}$, the qs are primes $\equiv 3$ or $7 \pmod{20}$, and the us are other primes. Show that if $d_1 + d_2 + \cdots + d_s$ and all the es are even, then n is represented by $X^2 + 5Y^2$.

8. Let $p \equiv 1 \pmod{12}$ be prime.

i. Show that $p = a^2 + b^2 = t^2 + 3u^2$, where a, b, t, u are positive integers, a and t are odd, and b and u are even.

ii. Prove that $(-3)^{(p-1)/4} \equiv (-1)^{(t-1)/2}(2/p)(t/3) \pmod{p}$.
(*Hint:* Show that $(t/p) = (-1)^{(t-1)/2}(t/3)$ and $(u/p) = (2/p)$.)

iii. Show that $3|a$ or $3|b$ but not both.

iv. Observe that $(t+b)(t-b) = a^2 - 3u^2$.

v. Suppose that $3|b$. Show that $3 \nmid t+b$ and verify that $(t/3) = ((t+b)/3) = (-1)^{(t-1)/2}(2/p)$. Hence $(-3)^{(p-1)/4} \equiv 1 \pmod{p}$.

vi. Suppose that $3|a$. Show that $-(t/3) = ((t \pm b)/3) = (-1)^{(t-1)/2}(2/p)$, where the sign is chosen so that the Legendre symbol $((t \pm b)/3)$ is defined, i.e., such that $3 \nmid t \pm b$. Hence $(-3)^{(p-1)/4} \equiv -1 \pmod{p}$.

vii. Prove that $X^2 + 36Y^2$ represents p if and only if $X^4 \equiv -3 \pmod{p}$ has a solution, and that $9X^2 + 4Y^2$ represents p if and only if $X^4 \equiv -3 \pmod{p}$ has no solution.

CHAPTER 4
Indefinite Forms

1. Introduction

In this chapter we solve an important class of Diophantine equations.

Theorem 1.1. Let $f(X, Y) = aX^2 + bXY + cY^2$ be an integral binary quadratic form and let $m \in \mathbb{Z}$. There is a finite algorithm to find all solutions $(x, y) \in \mathbb{Z}^2$ of the equation $f(x, y) = m$.

It turns out that Theorem 1.1 is most interesting if the discriminant of $f(X, Y) = aX^2 + bXY + cY^2$ is positive and nonsquare. Then the set of integral solutions of the equation $f(X, Y) = m$ for $m \neq 0$ is infinite if nonempty, so its members can not be listed. It is natural to try to impose a finitely generated algebraic structure on the set of solutions, then to list a set of generators. That is in fact what we will do. The most important case is the Pell equation $X^2 - DY^2 = 1$.

Theorem 1.2. Let D be a nonsquare positive integer.

i. The set $\mathcal{P}\!ell(4D) = \{(x, y) \in \mathbb{Z}^2 | x^2 - Dy^2 = 1\}$ is infinite.
ii. The binary operation $(x, y) \cdot (u, v) = (xu + Dyv, xv + yu)$ is a group law on $\mathcal{P}\!ell(4D)$ for which $\mathcal{P}\!ell(4D) \simeq \{\pm 1\} \times \mathbb{Z}$.

It will turn out that if the discriminant of f equals $4D$, then the group $\mathcal{P}\!ell(4D)$ acts on the set of integral solutions of the equation $f(X, Y) = m$ and the number of orbits is finite. The algorithm of Theorem 1.1 will produce a list containing one solution in each orbit.

We discuss the simple case $X^2 - 2Y^2 = 1$ in Section 2, then prove Theorem 1.2 in Section 3.

Theorem 1.1 is proved in Section 4 by the derivation of a straightforward algorithm to solve the equation $f(X, Y) = m$.

The set $\mathcal{P}ell(4D)$ can of course be shown to be a group by direct verification of the group axioms. But one can search for naturally occurring groups with which it is in bijection. We discuss two such groups. In Section 3 we show that $\mathcal{P}ell(4D)$ is isomorphic to the group of norm 1 units of the ring $\mathbb{Z}[\sqrt{D}]$. In Section 5 we prove that $\mathcal{P}ell(4D)$ is isomorphic to the group of proper automorphisms of any primitive form of discriminant $4D$.

The reduction theory described in Section 6 is a detailed study of the proper equivalence relation for forms of positive nonsquare discriminant. Proofs will occupy Sections 7 and 8.

The reduction theory is brought to bear on the automorphism group of a form in Section 9. This provides a second proof of Theorem 1.2.

2. The Square Root of 2

It must have been a research problem around 500BC to solve the equation $z^2 = 2$. The solution gives the length of the diagonal of a 1×1 square. The momentus discovery by the Greeks (Pythagoras?) that there is no rational number solution z amounted to the discovery of irrational numbers.

A rational number z is the quotient of two integers: $z = y/x$, where $x, y \in \mathbb{Z}$ and $x \neq 0$. So the Greeks' discovery can be presented as follows.

Theorem 2.1. Let $x, y \in \mathbb{Z}$ satisfy the equation: $y^2 = 2x^2$.
Then $x = y = 0$.

Proof. This is *the* classic descent. With x and y such that $y^2 = 2x^2$, we prove by induction that $x_n = x/2^n$ and $y_n = y/2^n$ are integers for every integer $n \geq 0$. Indeed, suppose that $x_n, y_n \in \mathbb{Z}$. We have $y_n^2 = 2x_n^2$, from which it is clear that y_n is even. Thus $2x_n^2$ is a multiple of 4, and so x_n is also even. Hence $x_{n+1}, y_{n+1} \in \mathbb{Z}$. But $|x_n| < 1$ for sufficiently large n, which implies $x_n = 0$. The result follows. ■

Failing to find a rational number that equals $\sqrt{2}$, we might search for rationals that *approximate* $\sqrt{2}$. It seems natural to look for integral points on the hyperbola $y^2 = 2x^2 + 1$ which is asymptotic to the line $y = \sqrt{2}\,x$. The situation is better immediately because there are integer points on the hyperbola, for instance $(x, y) = (2, 3)$ or $(13860, 19601)$.

The ancients knew how to "multiply" solutions of the equation $y^2 - 2x^2 = 1$: If $(x, y) = (u, v)$, (U, V) are two such solutions, then their "product"

$$(u, v) \cdot (U, V) = (uV + vU, 2uU + vV) \tag{2.2}$$

is a third. This follows from the key identity

$$(2uU + vV)^2 - 2(uV + vU)^2 = (v^2 - 2u^2)(V^2 - 2U^2). \tag{2.3}$$

Starting with one integer point (x, y) on the hyperbola in the first quadrant, we can find infinitely many others by taking "powers." For example, let $(x_1, y_1) = (2, 3)$ and let $(x_n, y_n) = (2, 3)(x_{n-1}, y_{n-1})$ for $n > 1$. We will write $(x_n, y_n) = (2, 3)^n$. We have the equality $y_n^2 - 2x_n^2 = 1$ for all $n \geq 1$ and, as is easily seen, the inequalities $x_1 < x_2 < x_3 < \cdots$ and $y_1 < y_2 < \cdots$.

Two interesting formulas for (x_n, y_n) are easily proved by finite induction:

$$\begin{pmatrix} x_n \\ y_n \end{pmatrix} = \begin{pmatrix} 3 & 2 \\ 4 & 3 \end{pmatrix}^n \begin{pmatrix} 0 \\ 1 \end{pmatrix} \quad \text{for } n \geq 1, \tag{2.4}$$

$$x_n\sqrt{2} + y_n = (2\sqrt{2} + 3)^n \quad \text{for } n \geq 1. \tag{2.5}$$

Table of solutions $y_n^2 - 2x_n^2 = 1$:

n	x_n	y_n	y_n/x_n
1	2	3	1.5
2	12	17	1.416...
3	70	99	1.41428...
4	408	577	1.414215...
5	2378	3363	1.4142136...
6	13860	19601	1.414213564...
7	80782	114243	1.4142135624...
8	470832	665857	1.414213562374...
9	2744210	3880899	1.4142135623731...
10	15994428	22619537	1.414213562373096...
			$\sqrt{2}$ = 1.4142135623730950...

Aside from its theoretical value, the power of (2.2) to generate quickly tables of integer solutions of the equation $y^2 - 2x^2 = 1$ like the one shown is really quite extraordinary.

When an equation has an infinite number of solutions we can not list them all. So what does it mean to "find all the solutions?" Usually it just means to find a pretty description of the set of solutions. Such a description is available for the equation $y^2 - 2x^2 = 1$.

Theorem 2.6. Let u, v be positive integers such that $v^2 - 2u^2 = 1$.

Then there is an integer $n \geq 1$ such that $(u, v) = (x_n, y_n)$. (In other words, $(u, v) = (2, 3)^n$ for some $n \geq 1$.)

Proof. The proof is by induction on u.

Let u, v be positive integers such that $v^2 - 2u^2 = 1$. Clearly $u = 1$ is impossible and $u = 2$ implies that $(u, v) = (2, 3) = (x_1, y_1)$. So suppose that $u > 2$. Let $(u_1, v_1) = (-2, 3) \cdot (u, v) = (3u - 2v, 3v - 4u)$, so that $(u, v) = (2, 3)(u_1, v_1)$ and $v_1^2 - 2u_1^2 = 1$. From $v/u > \sqrt{2}$ it follows that $u_1 < u$ and that $v_1 > 0$. Since $u > 2$, we can compute that $v = \sqrt{2u^2 + 1} = \sqrt{(3u/2)^2 + 1 - (u/2)^2} < 3u/2$, which means that $u_1 > 0$. Hence an inductive hypothesis applies to (u_1, v_1). We conclude that $(u_1, v_1) = (x_n, y_n)$ for some $n \geq 1$, and therefore $(u, v) = (x_{n+1}, y_{n+1})$. ∎

It is easy to bound the error with which y_n/x_n estimates $\sqrt{2}$. We find, for example, that the estimate with $n = 10$ is accurate to 14 decimal places, which is better than that provided by a hand-held calculator.

Proposition 2.7.

i.
$$\left|\frac{y_n}{x_n} - \sqrt{2}\right| < \frac{1}{2\sqrt{2}\,x_n^2} < 10^{-(3n-1)/2} \quad \text{for } n \geq 1.$$

ii. Let x, y be positive integers such that $|y/x - \sqrt{2}| < 1/2x^2$.

Then $y^2 - 2x^2 = \pm 1$.

Proof. i. Let $x, y > 0$ satisfy $y^2 - 2x^2 = 1$. Then

$$\left|\frac{y}{x} - \sqrt{2}\right| = \frac{y - x\sqrt{2}}{x} = \frac{1}{x(y + x\sqrt{2})} < \frac{1}{x(x\sqrt{2} + x\sqrt{2})} = \frac{1}{2\sqrt{2}\,x^2}.$$

Also

$$x_n = 3x_{n-1} + 2y_{n-1} > (3 + 2\sqrt{2})x_{n-1} > \cdots > (3 + 2\sqrt{2})^{n-1} x_1 \quad \text{for } n > 1.$$

Therefore,

$$\left|\frac{y_n}{x_n} - \sqrt{2}\right| < \frac{1}{8\sqrt{2}\,(3 + 2\sqrt{2})^{2n-2}} < 1/10^{(3n-1)/2},$$

where we have made use of the inequality $(3 + 2\sqrt{2})^4 > 10^3$.

ii. Let x, y be positive integers such that $|y/x - \sqrt{2}| < 1/2x^2$. Then $|y - \sqrt{2}\,x| < 1/2x$ and $|y + \sqrt{2}\,x| < y + y + 1/2x$. Multiplying the two inequalities yields

$$|y^2 - 2x^2| < \frac{y}{x} + \frac{1}{4x^2} < \sqrt{2} + \frac{1}{2x^2} + \frac{1}{4x^2}.$$

If $x \geq 2$, then $|y^2 - 2x^2|$ is an integer less than 2, which by Theorem 2.1 can only be 1. If $x = 1$, then $y = 1$, which gives $y^2 - 2x^2 = -1$. The proposition is proved. ∎

Exercises

1. Prove formulas (2.4) and (2.5). Show that $x_n = 6x_{n-1} - x_{n-2}$ and that $y_n = 6y_{n-1} - y_{n-2}$ for $n \geq 3$.

2. Multiply out $(u\sqrt{2} + v)(U\sqrt{2} + V)$ and compare with formula (2.2).

3. Show that the product (2.2) is a commutative group law on the set $\{(x, y) \in \mathbb{Z}^2 | y^2 - 2x^2 = 1\}$. Prove that the group is isomorphic to $\{\pm 1\} \times \mathbb{Z}$.

4. (Converse to Proposition 2.7ii.) Let x, y be positive integers such that $|y^2 - 2x^2| = 1$. Prove that $|y/x - \sqrt{2}| < 1/2x^2$.

5. **i.** Show that rule (2.2) defines a product on the set of pairs of integers (x, y) such that $|y^2 - 2x^2| = 1$.

ii. Let u, v be positive integers such that $|v^2 - 2u^2| = 1$. Prove that $(u, v) = (1, 1)^n$ for some $n \geq 1$.

6. Find all positive integer solutions x, y of the equation $y^2 - 3x^2 = 1$. Discuss the approximation of $\sqrt{3}$ by rational numbers.

3. The Pell Equation

Definition 3.1. Define the *Pell form* f_Δ for a nonzero discriminant Δ by the equation

$$f_\Delta(x, y) = \begin{cases} x^2 - \dfrac{\Delta}{4} y^2 & \text{if } \Delta \equiv 0 \pmod 4 \\[2ex] x^2 + xy - \dfrac{(\Delta - 1)}{4} y^2 & \text{if } \Delta \equiv 1 \pmod 4 \end{cases}.$$

The *Pell equation* is the equation $f_\Delta(x, y) = 1$. The *negative Pell equation* is the equation $f_\Delta(x, y) = -1$. Define $\mathcal{P}ell(\Delta) = \{(x, y) \in \mathbb{Z}^2 | f_\Delta(x, y) = 1\}$. Define $\mathcal{P}ell^{\pm}(\Delta) = \{(x, y) \in \mathbb{Z}^2 | f_\Delta(x, y) = \pm 1\}$.

Note that f_Δ is an integral binary form of discriminant Δ.

In this section we will find all solutions $(x, y) \in \mathbb{Z}^2$ of the Pell equation. It turns out that the solution set $\mathcal{P}ell(\Delta)$ can be made into a commutative group in a natural way. (The binary operation is given by (3.14).) We will determine the isomorphism class of the group. As will be seen in this chapter, the Pell equation has an important distinguished role in the theory of binary quadratic forms.

It is easy to see that the two trivial solutions $(1, 0)$ and $(-1, 0)$ are the only solutions $(x, y) \in \mathbb{Z}^2$ of the Pell equation if the discriminant Δ is the square of an integer or if $\Delta < -4$ (exercise). There are 4 and 6 integral solutions in the cases $\Delta = -4$ and $\Delta = -3$. $\mathcal{P}ell(\Delta)$ is a finite cyclic group in all these cases.

The facts are more interesting for a positive discriminant Δ that is not the square of an integer. The Pell equation with $\Delta = 8$, which is the equation $x^2 - 2y^2 = 1$ discussed in Section 2, is typical. The solution set is infinite but can be generated, in a sense to be made precise, by a single nontrivial solution. The hardest thing to prove is the existence of a nontrivial solution.

Section 3 is devoted to the Pell equation with positive nonsquare discriminant Δ. We begin by generalizing the theorem that $\sqrt{2}$ is irrational.

Proposition 3.2. Let $D \in \mathbb{Z}$. If $\sqrt{D} \notin \mathbb{Z}$, then $\sqrt{D} \notin \mathbb{Q}$.

Proof. If $\sqrt{D} \in \mathbb{Q}$, then there exist $a, b \in \mathbb{Z}$ with $b > 0$ such that $a^2 = Db^2$. We prove by induction on b that $b | a$. This is trivial if $b = 1$. If $b > 1$, let p be a prime divisor of b. Clearly $p | a^2$ and hence $p | a$. Since $(a/p)^2 = D(b/p)^2$, an inductive hypothesis shows that b/p divides a/p, whence $b | a$. It follows that $D = (a/b)^2 \in \mathbb{Z}$. ■

Thus it is the same to say that an integer is not the square of an integer and that it is not the square of a rational number. We will simply say that such an integer is *nonsquare*.

Theorem 3.3. Let $D \in \mathbb{Z}$ be a positive nonsquare. There exist $(x, y) \in \mathbb{Z}^2$ such that $x^2 - Dy^2 = 1$ and $y \neq 0$.

We need a lemma.

Lemma 3.4. Let $D \in \mathbb{Z}$ be a positive nonsquare and let m be a positive integer. Then there exist $(x, y) \in \mathbb{Z}^2$ such that $0 < y \le m$, $|x - y\sqrt{D}| < 1/m$, and $|x^2 - Dy^2| < 2\sqrt{D} + 1$.

Proof. By the pigeon-hole principle. Let $x_i = [i\sqrt{D}]$ for $i = 0, 1, \ldots, m$. Each of the $m + 1$ real numbers $i\sqrt{D} - x_i$ belongs to one of the m open intervals $(k/m, (k+1)/m)$, $0 \le k \le m - 1$. Hence there exist integers i and j such that $0 \le i < j \le m$ and $|(i\sqrt{D} - x_i) - (j\sqrt{D} - x_j)| < 1/m$. Taking $x = x_j - x_i$ and $y = j - i$, we get $0 < y \le m$ and $|x - y\sqrt{D}| < 1/m$. Multiplying the two inequalities $|x - y\sqrt{D}| < 1/m$ and $|x + y\sqrt{D}| = |(x - y\sqrt{D}) + 2y\sqrt{D}| < 1/m + 2m\sqrt{D}$ reveals that $|x^2 - Dy^2| < 1/m^2 + 2\sqrt{D} \le 2\sqrt{D} + 1$. ∎

Proof of Theorem 3.3. For each positive integer m there is a pair $(x, y) \in \mathbb{Z}^2$ as in Lemma 3.4. Because $\sqrt{D}$ is irrational and $y \ne 0$, one pair (x, y) can satisfy the inequality $|x - y\sqrt{D}| < 1/m$ for only a finite number of m. Hence there is an infinite number of distinct pairs $(x, y) \in \mathbb{Z}^2$ such that $|x^2 - Dy^2| < 2\sqrt{D} + 1$.

By the pigeon-hole principle there must exist $k \in \mathbb{Z}$, $0 < |k| < 2\sqrt{D} + 1$, such that $x^2 - Dy^2 = k$ has infinitely many solutions $(x, y) \in \mathbb{Z}^2$. Each such solution belongs to one of only k^2 congruence classes mod k, so by the pigeon-hole principle we can select a congruence class containing infinitely many of them. Thus there exist $(x_1, y_1), (x_2, y_2) \in \mathbb{Z}^2$ such that (3.5a, b, c) hold:

$$x_1^2 - Dy_1^2 = x_2^2 - Dy_2^2 = k \ne 0, \tag{3.5a}$$

$$x_1 - x_2 \equiv y_1 - y_2 \equiv 0 \pmod{k}, \tag{3.5b}$$

$$(x_1, y_1) \ne \pm(x_2, y_2). \tag{3.5c}$$

Define $x, y \in \mathbb{Q}$ by the equation

$$x + y\sqrt{D} = \frac{x_1 + y_1\sqrt{D}}{x_2 + y_2\sqrt{D}} = \left(x_1 + y_1\sqrt{D}\right)\frac{x_2 - y_2\sqrt{D}}{k}.$$

We finish the proof of Theorem 3.3 by showing that $(x, y) \in \mathbb{Z}^2$, $x^2 - Dy^2 = 1$, and $y \ne 0$.

Note that $x - y\sqrt{D} = (x_1 - y_1\sqrt{D})/(x_2 - y_2\sqrt{D})$. It follows that

$$x^2 - Dy^2 = \left(x + y\sqrt{D}\right)\left(x - y\sqrt{D}\right) = \left(x_1^2 - Dy_1^2\right)/\left(x_2^2 - Dy_2^2\right) = 1.$$

To prove that x and y are integers, begin with the expressions $x = (x_1x_2 - Dy_1y_2)/k$ and $y = (x_2y_1 - x_1y_2)/k$. Then calculate, using (3.5a) and (3.5b), that $kx \equiv x_1^2 - Dy_1^2 \equiv 0 \pmod{k}$ and $ky \equiv x_1y_1 - x_1y_1 \equiv 0 \pmod{k}$.

Now suppose that $y = 0$. From $x^2 - Dy^2 = 1$ it would follow that $x = \pm 1$, which would contradict the fact (3.5c) that $x + y\sqrt{D} \neq \pm 1$. We conclude that $y \neq 0$. ■

Corollary 3.6. The Pell equation $f_\Delta(x, y) = 1$ has a solution in positive integers x and y for every positive nonsquare discriminant Δ.

Proof. If $\Delta \equiv 0 \pmod 4$, this is just Theorem 3.3 where $D = \Delta/4$.

If $\Delta \equiv 1 \pmod 4$, choose $u, v \in \mathbb{Z}$ with u and $v > 0$ such that $u^2 - \Delta v^2 = 1$. Then $f_\Delta(x, y) = 1$, where $x = u - v$ and $y = 2v$. ■

The set of integral solutions of the equation $x^2 - Dy^2 = 1$ for a nonsquare positive integer D can be easily analyzed by following closely the model in Section 2 for $D = 2$. All solutions with positive x and y are suitable powers of a minimal "fundamental" such solution, which exists by Theorem 3.3. We prefer, however, to develop this theory in another language, that of rings generated by irrational square roots that generalize the ring of Gaussian integers. Though not essential, this clarifies the mathematics by giving a conceptual basis for the group law on $\mathcal{P}ell^{\pm}(\Delta)$. We proceed to the necessary preliminary formalities.

Define $\sqrt{x}$ for $x \neq 0 \in \mathbb{R}$ in the standard way: If $x > 0$, then $\sqrt{x} > 0$; if $x < 0$, then $\operatorname{Im}\sqrt{x} > 0$, i.e., $\sqrt{x} = i\sqrt{|x|} \in \mathbb{C}$.

***Definition** 3.7.* For a nonsquare discriminant Δ let $\mathbb{Q}(\sqrt{\Delta}) = \{x + y\sqrt{\Delta} \mid x, y \in \mathbb{Q}\}$. We define *conjugation* σ and *norm* N on $\mathbb{Q}(\sqrt{\Delta})$ as follows: For $x, y \in \mathbb{Q}$ and $\alpha = x + y\sqrt{\Delta}$, set $\sigma(\alpha) = x - y\sqrt{\Delta} \in \mathbb{Q}(\sqrt{\Delta})$ and $N(\alpha) = \alpha \cdot \sigma(\alpha) = x^2 - \Delta y^2 \in \mathbb{Q}$.

Note that the definition of the conjugation σ depends on the fact (Proposition 3.2) that $\sqrt{\Delta}$ is irrational for nonsquare discriminants Δ. In order to define $\sigma(\alpha)$ for $\alpha \in \mathbb{Q}(\sqrt{\Delta})$ in terms of an expression $\alpha = x + y\sqrt{\Delta}$, where $x, y \in \mathbb{Q}$, we must know that α has a *unique* such expression.

If $\Delta > 0$, then $\mathbb{Q}(\sqrt{\Delta}) \subset \mathbb{R}$. In all cases $\mathbb{Q}(\sqrt{\Delta}) \subset \mathbb{C}$. If $\Delta < 0$, then σ is just complex conjugation.

The elementary properties of $\mathbb{Q}(\sqrt{\Delta})$ that we will need are summarized in the next lemma.

Lemma 3.8. Let Δ be a nonsquare discriminant.

i. $\mathbb{Q}(\sqrt{\Delta})$ is a field.
ii. $\sigma\colon \mathbb{Q}(\sqrt{\Delta}) \to \mathbb{Q}(\sqrt{\Delta})$ is an isomorphism of fields.
iii. $N(\alpha\beta) = N(\alpha)N(\beta)$ for all $\alpha, \beta \in \mathbb{Q}(\sqrt{\Delta})$.
iv. $N(\alpha) = 0$ if and only if $\alpha = 0$, for all $\alpha \in \mathbb{Q}(\sqrt{\Delta})$.

Proof. iv holds because Δ is not a rational square. $\mathbb{Q}(\sqrt{\Delta})$ is clearly a ring. The calculation $\alpha(\sigma(\alpha)/N(\alpha)) = 1$ for $\alpha \neq 0 \in \mathbb{Q}(\sqrt{\Delta})$ shows that α is invertible and hence that $\mathbb{Q}(\sqrt{\Delta})$ is a field. A quick computation shows that σ preserves sums and products. Since σ is its own inverse, σ is a bijection and is hence a field isomorphism. Multiplicativity of N follows from multiplicativity of σ: $N(\alpha\beta) = \alpha\beta\sigma(\alpha\beta) = \alpha\sigma(\alpha)\beta\sigma(\beta) = N(\alpha)N(\beta)$. ∎

The multiplicativity of the norm map, Lemma 3.8iii, generalizes the identity (2.3). To see how, let $\alpha_i = x_i + y_i\sqrt{\Delta}$, where $x_i.\ y_i \in \mathbb{Q}$. The equation $N(\alpha_1\alpha_2) = N(\alpha_1)N(\alpha_2)$ is the identity

$$(x_1x_2 + \Delta y_1y_2)^2 - \Delta(x_1y_2 + y_1x_2)^2 = (x_1^2 - \Delta y_1^2)(x_2^2 - \Delta y_2^2).$$

The focus for the rest of Section 3 will be on an important subring $\mathcal{O}_\Delta$ of $\mathbb{Q}(\sqrt{\Delta})$.

Definition 3.9. The Δ-order $\mathcal{O}_\Delta$ is defined for nonsquare discriminants Δ to be the ring $\mathcal{O}_\Delta = \{x + y\rho_\Delta | x, y \in \mathbb{Z}\}$ where

$$\rho_\Delta = \begin{cases} \sqrt{\Delta/4} & \text{if } \Delta \equiv 0 \pmod 4 \\ (1 + \sqrt{\Delta})/2 & \text{if } \Delta \equiv 1 \pmod 4 \end{cases}.$$

A trivial check (exercise) shows that $\mathcal{O}_\Delta$ is actually a subring of $\mathbb{Q}(\sqrt{\Delta})$ as claimed. The congruence condition on Δ enters into the verification that $\mathcal{O}_\Delta$ is closed under multiplication. Note that $\mathcal{O}_{-4}$ is nothing but the ring of Gaussian integers.

It is the norm map that yokes the rings $\mathcal{O}_\Delta$ to the Diophantine equations we are aiming to study. The basic calculation:

$$N(x + y\rho_\Delta) = f_\Delta(x, y) \quad \text{for all } x, y \in \mathbb{Q}. \tag{3.10}$$

Thus solving the Pell equation means finding the elements of $\mathcal{O}_\Delta$ of norm $+1$. Since these elements are in a ring they can be multiplied, which explains the earlier product rule (2.2) for combining solutions of the equation $x^2 - 2y^2 = 1$.

Lemma 3.11. Let Δ be a nonsquare discriminant and let $\alpha \in \mathcal{O}_\Delta$. Then α is a unit in $\mathcal{O}_\Delta$ if and only if $N(\alpha) = \pm 1$.

Proof. If α is a unit in $\mathcal{O}_\Delta$, then there exists $\beta \in \mathcal{O}_\Delta$ such that $\alpha\beta = 1$. It follows that $N(\alpha)N(\beta) = 1$. Since all elements of $\mathcal{O}_\Delta$ have integral norm, we can conclude that $N(\alpha) = \pm 1$. Conversely, if $N\alpha = \pm 1$ for $\alpha \in \mathcal{O}_\Delta$, then $\alpha^{-1} = \sigma(\alpha)/N(\alpha)$ belongs to $\mathcal{O}_\Delta$ and so α is a unit. ∎

We have discovered, (3.10) and Lemma 3.11, that the problem of solving the positive and negative Pell equations is the same as the problem of determining the group of units of the ring $\mathcal{O}_\Delta$. It is natural to make use of the group structure in describing the set of solutions of the equations. Some notation for the relevant groups of units will facilitate the discussion.

Notation 3.12. The unit group $\mathcal{O}_\Delta^\times$ is defined for nonsquare discriminants Δ to be the group of units of the ring $\mathcal{O}_\Delta$.

Define $\mathcal{O}_{\Delta,1}^\times = \{\alpha \in \mathcal{O}_\Delta^\times | N(\alpha) = +1\}$, the subgroup of units for norm $+1$.

For $\Delta > 0$, define $\mathcal{O}_{\Delta,+}^\times = \{\alpha \in \mathcal{O}_\Delta^\times | \alpha > 0\}$, the subgroup of positive units.

By (3.10) and Lemma 3.11 there is a bijection $\psi\colon \mathcal{P}ell^{\pm}(\Delta) \to \mathcal{O}_\Delta^\times$ given by $\psi(x, y) = x + y\rho_\Delta$. Since it is in bijection with a commutative group, $\mathcal{P}ell^{\pm}(\Delta)$ itself is a group for every nonsquare discriminant Δ. We simply use ψ to transport the group law from $\mathcal{O}_\Delta^\times$, so that by definition $a \cdot b = \psi^{-1}(\psi(a)\psi(b))$ for every $a, b \in \mathcal{P}ell^{\pm}(\Delta)$. In other words, the product $(u, v) \cdot (U, V)$ of two elements $(u, v), (U, V) \in \mathcal{P}ell^{\pm}(\Delta)$ is defined by the rule

$$(u, v) \cdot (U, V) = (x, y), \quad \text{where } x + y\rho_\Delta = (u + v\rho_\Delta)(U + V\rho_\Delta). \tag{3.13}$$

It follows by a calculation (exercise) that

$$(u, v) \cdot (U, V) = \begin{cases} \left(uU + \dfrac{\Delta}{4}vV, uV + vU\right) & \text{if } \Delta \equiv 0 \pmod 4 \\ \left(uU + \dfrac{\Delta - 1}{4}vV, uV + vU + vV\right) & \text{if } \Delta \equiv 1 \pmod 4 \end{cases}. \tag{3.14}$$

The group structure on $\mathcal{P}ell^{\pm}(\Delta)$ has been defined just so that $\psi\colon \mathcal{P}ell^{\pm}(\Delta) \to \mathcal{O}_\Delta^\times$ is an isomorphism of groups. Restricting ψ to subgroups gives an isomorphism $\mathcal{P}ell(\Delta) \simeq \mathcal{O}_{\Delta,1}^\times$. Solving the Pell equation will mean determining the groups $\mathcal{P}ell(\Delta)$.

There is a construction of $\mathcal{O}_\Delta$ for positive square discriminants Δ given in the exercises. This leads to a group structure on $\mathcal{P}ell^{\pm}(\Delta)$ with binary operation given by (3.14). Alternatively, it can of course be verified by tedious direct calculation that (3.14) is a group law on $\mathcal{P}ell^{\pm}(\Delta)$ with identity element $(1, 0)$ for all nonzero discriminants Δ. Thus the ring $\mathcal{O}_\Delta$ need not intervene in the development at all.

The explicit determination of the groups $\mathcal{P}ell(\Delta)$ for negative discriminants and for positive square discriminants is relegated to the exercises. For the rest of this section we will deal exclusively with *positive nonsquare* discriminants Δ.

We can now state and prove the main theorem of Section 3.

Theorem 3.15. $\mathcal{O}_{\Delta,+}^{\times} \simeq \mathbb{Z}$ for every positive nonsquare discriminant Δ.

Proof. Let Δ be a nonsquare positive discriminant.

$\mathcal{O}_{\Delta,+}^{\times}$ is a subgroup of the multiplicative group of positive real numbers. First note that $\mathcal{O}_{\Delta,+}^{\times}$ contains units greater than 1. Indeed, by Corollary 3.6 there exist positive integers x and y such that $f_\Delta(x, y) = 1$. Then $\alpha = x + y\rho_\Delta \in \mathcal{O}_{\Delta,+}^{\times}$ and $\alpha > 1$. We prove next that $\mathcal{O}_{\Delta,+}^{\times}$ contains a minimal element greater than 1.

Let $\alpha \in \mathcal{O}_{\Delta,+}^{\times}$. Then α is a root of the polynomial $(X - \alpha)(X - \sigma(\alpha)) = X^2 - mX \pm 1$, where $m = \alpha + \sigma(\alpha) \in \mathbb{Z}$. If $\alpha > 1$, then $|m| \leq \alpha + |\sigma(\alpha)| < \alpha + 1$. Let $B > 1$. Every $\alpha \in \mathcal{O}_{\Delta,+}^{\times}$ such that $1 < \alpha \leq B$ is a root of one of the polynomials $X^2 - mX \pm 1$ such that $m \in \mathbb{Z}$ and $|m| < B + 1$. Since the roots of a finite number of polynomials are finite in number, we can conclude that $Y = \{\alpha \in \mathcal{O}_{\Delta,+}^{\times} | 1 < \alpha \leq B\}$ is a finite set. If B is large enough, then Y is nonempty and so must contain a minimal element ϵ. Then ϵ is clearly minimal in $\{\alpha \in \mathcal{O}_{\Delta,+}^{\times} | \alpha > 1\}$.

Now let $\alpha \in \mathcal{O}_{\Delta,+}^{\times}$. There exists $n \in \mathbb{Z}$ such that $\epsilon^n \leq \alpha < \epsilon^{n+1}$. Then $1 \leq \alpha/\epsilon^n < \epsilon$. The minimality of ϵ among units of $\mathcal{O}_\Delta$ greater than 1 implies that $\alpha/\epsilon^n = 1$, so $\alpha = \epsilon^n$. Therefore the homomorphism $\phi: \mathbb{Z} \to \mathcal{O}_{\Delta,+}^{\times}$, $\phi(n) = \epsilon^n$, is a group isomorphism. ∎

Corollary 3.16. Let Δ be a positive nonsquare discriminant.

Let ϵ_Δ be the smallest unit of $\mathcal{O}_\Delta$ that is greater than 1 and let

$$\tau_\Delta = \begin{Bmatrix} \epsilon_\Delta & \text{if } N\epsilon_\Delta = +1 \\ \epsilon_\Delta^2 & \text{if } N\epsilon_\Delta = -1 \end{Bmatrix}.$$

Then $\mathcal{P}ell^{\pm}(\Delta) \simeq \mathcal{O}_\Delta^{\times} = \{\pm\epsilon_\Delta^n | n \in \mathbb{Z}\} \simeq \{\pm 1\} \times \mathbb{Z}$ and $\mathcal{P}ell(\Delta) \simeq \mathcal{O}_{\Delta,1}^{\times} = \{\pm\tau_\Delta^n | n \in \mathbb{Z}\} \simeq \{\pm 1\} \times \mathbb{Z}$.

Proof. Since $-1 \in \mathcal{O}_{\Delta,1}^{\times}$, every $\alpha \in \mathcal{O}_\Delta^{\times}$ can be written in the form $\alpha = \operatorname{sign}(\alpha)|\alpha|$, where $|\alpha| \in \mathcal{O}_{\Delta,+}^{\times}$ and $N(|\alpha|) = N(\alpha)$. ∎

Definition 3.17. The *fundamental unit* ϵ_Δ is defined for a positive nonsquare discriminant Δ to be the smallest unit of $\mathcal{O}_\Delta$ that is greater than 1.

The message of Corollary 3.16 is that all integral solutions of the Pell equation $f_\Delta(x, y) = 1$ for a positive nonsquare discriminant Δ can be found from knowledge of the fundamental unit ϵ_Δ. In the remainder of Section 3 we spell this out for the case of even discriminant Δ, including a crude method to find ϵ_Δ.

Theorem 3.18. Let D be a positive nonsquare integer, so that $f_{4D}(x, y) = x^2 - Dy^2$.

i. Let y be the smallest positive integer such that one of $Dy^2 + 1$ or $Dy^2 - 1$ is a square and let x be the positive integer square root. Then $\epsilon_{4d} = x + y\sqrt{D}$.

ii. Define $(x_n, y_n) \in \mathbb{Z}^2$ by the equation $x_n + y_n\sqrt{D} = \epsilon_{4D}^n$, for all $n \in \mathbb{Z}$. Then $\mathcal{P}ell^{\pm}(4D) = \{\pm(x_n, y_n) | n \in \mathbb{Z}\}$ and $x_n^2 - Dy_n^2 = (N\epsilon_{4D})^n = (x_1^2 - Dy_1^2)^n$.

Proof. ii. This is just a restatement of Corollary 3.16.

i. Let $\alpha = x + y\sqrt{D}$, where $(x, y) \in \mathcal{P}ell^{\pm}(4D)$. We observe that $\alpha > 1$ if and only if $x > 0$ and $y > 0$. Indeed, if $\alpha > 1$, then $|\sigma(\alpha)| = 1/\alpha < 1$ and so $\alpha > \pm\sigma(\alpha)$. The two equations $x + y\sqrt{D} > x - y\sqrt{D}$ and $x + y\sqrt{D} > -x + y\sqrt{D}$ show that $x > 0$ and $y > 0$. The converse is trivial.

Now let (x_n, y_n) be as in Theorem 3.18ii, i.e., $x_n + y_n\sqrt{D} = \epsilon_{4D}^n$. Since $\epsilon_{4D} > 1$, we have $x_n > 0$ and $y_n > 0$ for $n > 0$. Thus $y_{n+1} = x_1y_n + y_1x_n > x_1y_n \geq y_n$ for all $n \geq 1$. Hence we have the inequalities $0 < y_1 < y_2 < y_3 < \cdots$. So to find the fundamental unit ϵ_{4D}, look for the smallest positive integer y for which there is a positive integer x satisfying one of the two equations $x^2 - Dy^2 = \pm 1$, then $\epsilon_{4D} = x + y\sqrt{D}$. This is what we set out to prove. ■

Consider the example $\Delta = 102 = 4 \cdot 26$. Since $26 \cdot 1^2 - 1 = 5^2$, we can write $\epsilon_{102} = 5 + \sqrt{26}$. Set $x_n + y_n\sqrt{26} = (5 + \sqrt{26})^n$. We have $\mathcal{P}ell^{\pm}(102) = \{\pm(x_n, y_n) | n \in \mathbb{Z}\}$, and, since $N\epsilon_{102} = -1$, $\mathcal{P}ell(\Delta) = \{\pm(x_{2n}, y_{2n}) | n \in \mathbb{Z}\}$. It is easy to prove that

$$\begin{pmatrix} x_n \\ y_n \end{pmatrix} = \begin{pmatrix} 5 & 26 \\ 1 & 5 \end{pmatrix}^n \begin{pmatrix} 1 \\ 0 \end{pmatrix}.$$

If we want all solutions of $x^2 - 26y^2 = 1$ in *positive* integers x and y, we just take (x_{2n}, y_{2n}) for *positive* integers n. We can begin a table of solutions of $x^2 - 26y^2 = 1$:

n	2	4	6
(x_n, y_n)	(51, 10)	(5201, 1020)	(530451, 104030)

It must be observed that the algorithm Theorem 3.18i to find the fundamental unit $\epsilon_{4D} = x + y\sqrt{D}$ is horribly slow if y should happen to be very large, as it can be even for quite small D. A much faster method to find ϵ_D will be presented in Section 9.

Exercises

1. Let $D \in \mathbb{Z}$ and let n be a positive integer. Prove that if $\sqrt[n]{D} \notin \mathbb{Z}$, then $\sqrt[n]{D} \notin \mathbb{Q}$.

2. Find all integer solutions of $x^2 - Dy^2 = 1$ for $D = 7, 8, 10, 11$, and 23.

3. Find all integral solutions of the equation $x^2 + 6xy + 7y^2 = 1$.

4. Find the five smallest positive integers x such that $(x+1)^3 - x^3$ is the square of an integer.

5. **i.** Let $M \in \mathbf{GL}_2(\mathbb{R})$, let $a, b \in \mathbb{R}$, and let

$$\begin{pmatrix} x_n \\ y_n \end{pmatrix} = M^n \begin{pmatrix} a \\ b \end{pmatrix}$$

for $n \in \mathbb{Z}$. Prove that $x_n = \operatorname{tr}(M)x_{n-1} - \det(M)x_{n-2}$ and $y_n = \operatorname{tr}(M)y_{n-1} - \det(M)y_{n-2}$ for all $n \in \mathbb{Z}$.

ii. Let Δ be a nonsquare positive discriminant and let $\epsilon_\Delta^n = x_n + y_n \rho_\Delta$ with $x_n, y_n \in \mathbb{Z}$ for all $n \in \mathbb{Z}$. Find $\alpha, \beta \in \mathbb{Z}$ such that $x_n = \alpha x_{n-1} - \beta x_{n-2}$ and $y_n = \alpha y_{n-1} - \beta y_{n-2}$ for all n.

6. Prove that $\mathcal{O}_\Delta$ is a ring for nonsquare discriminants Δ. Verify that (3.14) follows from (3.13).

7. Let $D \in \mathbb{Z}$. Prove that $x^2 - Dy^2 = -1$ has no integral solution if

i. D has a prime divisor congruent to 3 (mod 4), **ii.** $D \equiv 3 \pmod 4$, or **iii.** $D \equiv 0 \pmod 4$.

8. Let p be a prime, $p \equiv 1 \pmod 4$.

i. Let $(u, v) \in \mathbb{Z}^2$ be the solution of the equation $x^2 - py^2 = 1$ with $u > 0$, $v > 0$, and v minimal. Show that $u \equiv 1$, $v \equiv 0 \pmod 2$, and that $(u+1)/2 \cdot (u-1)/2 = p(v/2)^2$.

ii. Show that there exist $a, b \in \mathbb{Z}$ such that $(u-1)/2 = a^2$ and $(u+1)/2 = pb^2$.

iii. Prove that there exist $x, y \in \mathbb{Z}$ such that $x^2 - py^2 = -1$.

iv. Prove that there exist $x, y \in \mathbb{Z}$ such that $f_p(x, y) = -1$.

9. Let Δ be a nonzero discriminant. Verify by direct calculation that $f_\Delta((u,v)(U,V)) = f_\Delta(u,v) f_\Delta(U,V)$ for all $(u,v),(U,V) \in \mathbb{Q}^2$, where

$(u, v)(U, V)$ is defined by (3.14). Hence show that both $\mathcal{Pell}^{\pm}(\Delta)$ and $\mathcal{Pell}(\Delta)$ are closed under the binary operation given by formula (3.14). Show directly that (3.14) is a group law on $\mathcal{Pell}^{\pm}(\Delta)$.

10. Let D be a positive nonsquare integer. Let $(x, y) \in \mathbb{Z}^2$ be the solution of the equation $x^2 - Dy^2 = 1$ with $x > 0$, $y > 0$, and y minimal. Show by the method of proof of Theorem 2.6 that

$$\{(u, v) \in \mathbb{Z}^2 | u^2 - Dv^2 = 1, u > 0, v > 0\} = \{(x, y)^n | n > 0 \in \mathbb{Z}\}.$$

11. Let $A = \mathbb{Q} \times \mathbb{Q}$ be given a ring structure in the usual way: $(x, y) \cdot (X, Y) = (xX, yY)$. Define *conjugation* σ and *norm* N on A by the formulas $\sigma(x, y) = (y, x) \in A$ and $N(x, y) = xy \in \mathbb{Q}$. Prove that σ: $A \to A$ is a ring isomorphism and that $N(\alpha\beta) = N(\alpha)N(\beta)$ for $\alpha, \beta \in A$. Let m be a positive integer and let $\Delta = m^2$. Note that Δ is a discriminant. Let $\mathcal{O}_\Delta = \{(u, v) \in \mathbb{Z}^2 | u \equiv v \pmod m\}$. Prove that $\mathcal{O}_\Delta$ is a subring of A and determine its group of units. Let $\rho_\Delta = (m/2, -m/2)$ if m is even, $\rho_\Delta = ((1 + m)/2, (1 - m)/2)$ if m is odd. Show that $\mathcal{O}_\Delta = \{x(1,1) + y\rho_\Delta | (x, y) \in \mathbb{Z}^2\}$ and that $N(x(1,1) + y\rho_\Delta) = f_\Delta(x, y)$. Prove that ψ: $\mathcal{Pell}^{\pm}(\Delta) \to \mathcal{O}_\Delta^\times$, $\psi(x, y) = x(1,1) + y\rho_\Delta$, is a bijection. Show that the group law on $\mathcal{Pell}^{\pm}(\Delta)$ that is induced by ψ is given by (3.14).

12. Let Δ be a nonzero discriminant. Sometimes the equation $x^2 - \Delta y^2 = 4$ is called the Pell equation. Justify this terminology by proving that there is a bijection λ: $\mathcal{Pell}(\Delta) \to \{(x, y) \in \mathbb{Z}^2 | x^2 - \Delta y^2 = 4\}$ given by the formula

$$\lambda(u, v) = \begin{cases} (2u, v) & \text{if } \Delta \equiv 0 \pmod 4 \\ (2u + v, v) & \text{if } \Delta \equiv 1 \pmod 4 \end{cases}.$$

13. **i.** Show that $\mathcal{Pell}(\Delta) = \{\pm(1,0)\}$ if Δ is a positive square or if $\Delta < -4$.
ii. Show that $\mathcal{Pell}(-4) = \{\pm(1,0), \pm(0,1)\}$, a cyclic group of order 4.
iii. Show that $\mathcal{Pell}(-3) = \{\pm(1,0), \pm(0,1), \pm(1,-1)\}$, a cyclic group of order 6.

14. Let Δ be a nonzero discriminant. Show that $\mathcal{O}_{m^2\Delta} \subseteq \mathcal{O}_\Delta$ and that $\mathcal{O}_{m^2\Delta}^\times \subseteq \mathcal{O}_\Delta^\times$ for every $m \neq 0 \in \mathbb{Z}$. Show that if Δ is a positive nonsquare, then $\epsilon_{m^2\Delta}$ is a positive power of ϵ_Δ for every $m \neq 0 \in \mathbb{Z}$.

15. Let D be a nonsquare positive integer and let $(x_n, y_n) \in \mathbb{Z}^2$ be defined by the equation $\epsilon_{4D}^n = x_n + y_n\sqrt{D}$ for $n \in \mathbb{Z}$.
i. Prove that $\mathrm{GCD}(x_n, y_n) = 1$.

ii. Show that $\lim_{n\to\infty} x_n/y_n = \sqrt{D}$.

iii. Prove that $|x_n/y_n - \sqrt{D}| < 1/2\sqrt{D-1}\,y_n^2$ for all $n > 0$.

16. Let Δ be a nonsquare positive odd discriminant.

i. Let $\alpha = x + y\rho_\Delta \in \mathcal{O}_\Delta^\times$, with $(x, y) \in \mathbb{Z}^2$. Show that $\alpha > 1$ if and only if $x \geq 0$ and $y > 0$. ($x > 0$ and $y > 0$ if $\Delta \neq 5$.)

ii. Let $\epsilon_\Delta^n = x_n + y_n\rho_\Delta$, where $(x_n, y_n) \in \mathbb{Z}^2$, for $n \in \mathbb{Z}$. Prove that $0 \leq x_1 < x_2 \leq x_3 < \cdots < \cdots$ and $0 < y_1 \leq y_2 < y_3 < \cdots < \cdots$, where $x_1 = 0$, $x_2 = x_3$, and $y_1 = y_2$ only if $\Delta = 5$.

iii. Find ϵ_Δ for $\Delta = 5, 13, 17, 21, 29, 33$.

iv. Show that $\mathrm{GCD}(x_n, y_n) = 1$, that $\lim_{n\to\infty} x_n/y_n = (\sqrt{\Delta} - 1)/2$, and that $|x_n/y_n - (\sqrt{\Delta} - 1)/2| < 1/\sqrt{\Delta - 4}\,y_n^2$ for all $n > 0$.

17. Show that $\epsilon_5^n = F_{n-1} + F_n\rho_5$ for $n > 0$, where the F_i are the Fibonacci numbers: $F_0 = 0$, $F_1 = 1$, and $F_i = F_{i-1} + F_{i-2}$ for $i \geq 2$.

18. Find an infinite number of integral solutions of the equation $x^2 - 7y^2 = 2$.

19. Let d be an integer that is not a cube, let $\theta = \sqrt[3]{d} \in \mathbb{R}$, and let $g_d(x, y, z) = x^3 + dy^3 + d^2z^3 - 3dxyz$. There is a binary operation on the set $X_d = \{(x, y, z) \in \mathbb{Z}^3 | g_d(x, y, z) = 1\}$ given by the rule $(x, y, z)(u, v, w) = (X, Y, Z)$, where $X + Y\theta + Z\theta^2 = (x + y\theta + z\theta^2)(u + v\theta + w\theta^2)$. In this way X_d becomes a commutative group, in fact in natural isomorphism with a subgroup of index 2 of the group of units of the ring $\mathbb{Z}[\theta] = \{x + y\theta + z\theta^2 | (x, y, z) \in \mathbb{Z}^3\} \subset \mathbb{R}$. It can be shown that $X_d \simeq \mathbb{Z}$. It can be shown that X_2 is generated by $(1, 1, 1)$ and that X_6 is generated by $(1, -6, 3)$. Calculate the second and third powers of these generators, and thereby find more elements of X_2 and X_6.

4. $aX^2 + bXY + cY^2 = m$

Let $f(X, Y) = aX^2 + bXY + cY^2$ be an integral binary quadratic form and let $m \in \mathbb{Z}$. We will show that there is a finite algorithm to find all solutions $(x, y) \in \mathbb{Z}^2$ of the equation $f(x, y) = m$. But the algorithm to be presented is dreadfully slow. Another method will be sketched in the exercises for Section 6.

The nature of the problem depends on the discriminant Δ of f.

If Δ is a square, then it is easy to see (exercise) that f is a product of two linear forms with integer coefficients: $f(x, y) = (\alpha x + \beta y)(\gamma x + \delta y)$ where

$\alpha, \beta, \gamma, \delta \in \mathbb{Z}$ and $\Delta = \det\begin{pmatrix} \alpha & \beta \\ \gamma & \delta \end{pmatrix}^2$. Given a factorization $m = pq$, where $p, q \in \mathbb{Z}$, we can solve the system $\alpha X + \beta Y = p$, $\gamma X + \delta Y = q$ for $(X, Y) \in \mathbb{Z}^2$ by the method of Chapter 1, Section 4 (Euclidean algorithm). Doing this for all factorizations of m we will find all integral solutions of the original equation $f(X, Y) = m$. Factoring integers is hard, but it cannot be avoided if $\Delta > 0$, as is shown by the equation $xy = m$. If $\Delta = 0$, then no factoring is required since $f(X, Y) = A(\alpha x + \beta y)^2$, where $A, \alpha, \beta \in \mathbb{Z}$.

If $\Delta < 0$, then completing the square on f leads to bounds on the size of solutions of the equation $f(X, Y) = m$. The expression $f(x, y) = ((ax + (b/2)y)^2 + |\Delta/4|y^2)/a = (|\Delta/4|x^2 + ((b/2)x + cy)^2)/c = m$ implies that $|x| \leq |4cm/\Delta|^{1/2}$ and $|y| \leq |4am/\Delta|^{1/2}$. All integral solutions of the equation $f(X, Y) = m$ belong to the set $\mathscr{X}$ of pairs $(x, y) \in \mathbb{Z}^2$ satisfying these bounds. Since $\mathscr{X}$ is finite, it can be searched for solutions.

If Δ is a positive nonsquare the problem is more subtle, as for the Pell equation. The set of integral solutions of $f(X, Y) = m$ if nonempty is infinite, so it is natural to try to describe the solution set in terms of some algebraic structure. The main theorem of Section 4 is that the group $\mathscr{P}ell(\Delta)$ acts on the set $\mathscr{Y} = \{(x, y) \in \mathbb{Z}^2 | f(x, y) = m\}$, that the number of orbits is finite, and that a list containing exactly one element from each orbit in $\mathscr{Y}$ can be produced. Such a list amounts to a solution of the equation $f(X, Y) = m$.

For the remainder of Section 4 all discriminants Δ are supposed nonsquare.

We begin by generalizing the construction of the ring $\mathscr{O}_\Delta$ from the Pell form f_Δ. The process is motivated by the factorization

$$ax^2 + bxy + cy^2 = \left(xa + y\frac{b + \sqrt{\Delta}}{2}\right)\left(xa + y\frac{b - \sqrt{\Delta}}{2}\right) \Big/ a. \qquad (4.1)$$

Definition 4.2. The *module* M_f of an integral binary quadratic form $f = ax^2 + bxy + cy^2$ of nonsquare discriminant $\Delta = b^2 - 4ac$ is the $\mathscr{O}_\Delta$ module $M_f = \{xa + y(b + \sqrt{\Delta})/2 | x, y \in \mathbb{Z}\} \subset \mathbb{Q}(\sqrt{\Delta}) \subset \mathbb{C}$. Note that $M_{f_\Delta} = \mathscr{O}_\Delta$.

It must be checked that M_f is really an $\mathscr{O}_\Delta$ module. The key point is closure of M_f under scalar multiplication by elements of $\mathscr{O}_\Delta$. The necessary calculation is

$$(u + v\rho_\Delta)\left(xa + y\frac{b + \sqrt{\Delta}}{2}\right) = x'a + y'\frac{b + \sqrt{\Delta}}{2},$$

where

$$(x', y') = \begin{cases} (x, y)\begin{pmatrix} u - \frac{b}{2}v & av \\ -cv & u + \frac{b}{2}v \end{pmatrix} & \text{if } \Delta \equiv 0 \pmod 4 \\ (x, y)\begin{pmatrix} u + \frac{1-b}{2}v & av \\ -cv & u + \frac{1+b}{2}v \end{pmatrix} & \text{if } \Delta \equiv 1 \pmod 4 \end{cases}. \tag{4.3}$$

If $u, v, x, y \in \mathbb{Z}$ in (4.3), then also $x', y' \in \mathbb{Z}$, because $b \equiv \Delta \pmod 2$.

It is the norm map on M_f that interests us.

Proposition 4.4. Let $f = ax^2 + bxy + cy^2$ be an integral form of nonsquare discriminant Δ. The formula $\psi(x, y) = xa + y(b + \sqrt{\Delta})/2$ defines a bijection

$$\psi\colon \{(x, y) \in \mathbb{Z}^2 | f(x, y) = m\} \to \{\gamma \in M_f | N(\gamma) = am\}.$$

Proof. This is just a fancy way of saying that $N(xa + y(b + \sqrt{\Delta})/2) = af(x, y)$ for all $(x, y) \in \mathbb{Z}^2$, which follows from (4.1). ■

Because the norm N is multiplicative, the group $\mathcal{O}_{\Delta,1}^{\times}$ of units of norm $+1$ in $\mathcal{O}_{\Delta}$ acts by scalar multiplication on the set $\mathcal{X} = \{\gamma \in \mathcal{M}_f | N(\gamma) = am\}$: If $\alpha \in \mathcal{O}_{\Delta,1}^{\times}$ and $\gamma \in \mathcal{X}$, then $N(\alpha\gamma) = N(\gamma)$, which shows that $\alpha\gamma \in \mathcal{X}$. By Proposition 4.4, $\mathcal{O}_{\Delta,1}^{\times}$ also acts on $\mathcal{Y} = \{(x, y) \in \mathbb{Z}^2 | f(x, y) = m\}$: By definition $\alpha \cdot (x, y) = \psi^{-1}(\alpha \cdot \psi(x, y))$ for all $\alpha \in \mathcal{O}_{\Delta,1}^{\times}$ and $(x, y) \in \mathcal{Y}$. The action of $\mathcal{O}_{\Delta,1}^{\times}$ on $\mathcal{Y}$ is given explicitly by the formulas (4.3).

The action of $\mathcal{O}_{\Delta,1}^{\times}$ on the set $\mathcal{Y}$ of integral solutions of the equation $f(X, Y) = m$ is most interesting when the discriminant Δ of f is a positive nonsquare, because then the group $\mathcal{O}_{\Delta,1}^{\times}$ is infinite. The orbit of each solution will then be infinite, and so the set $\mathcal{Y}$ will be either empty or infinite. The principal result of Section 4 is that the number of $\mathcal{O}_{\Delta,1}^{\times}$ orbits in $\mathcal{Y}$ is finite and that a list containing exactly one element in each orbit can be presented. Since $\mathcal{O}_{\Delta,1}^{\times}$ can be explicitly determined, the set $\mathcal{Y}$ is satisfactorily described by the presentation of such a list, called a *set of representatives* of the orbits.

Theorem 4.5. Let $f(x, y) = ax^2 + bxy + cy^2$ be an integral form of nonsquare positive discriminant $\Delta = b^2 - 4ac$ and let $m \neq 0 \in \mathbb{Z}$.

Let $\tau = \tau_\Delta$ be the smallest unit of $\mathcal{O}^\times_{\Delta,1}$ that is greater than 1.

i. Every $\mathcal{O}^\times_{\Delta,1}$ orbit of integral solutions of the equation $f(X, Y) = m$ contains a solution $(x, y) \in \mathbb{Z}^2$ such that

$$0 \le y \le U = \begin{cases} \left|\dfrac{am\tau}{\Delta}\right|^{1/2}\left(1 - \dfrac{1}{\tau}\right) = \left|\dfrac{am}{\Delta}(\tau + \sigma(\tau) - 2)\right|^{1/2} & \text{if } am > 0 \\ \left|\dfrac{am\tau}{\Delta}\right|^{1/2}\left(1 + \dfrac{1}{\tau}\right) = \left|\dfrac{am}{\Delta}(\tau + \sigma(\tau) + 2)\right|^{1/2} & \text{if } am < 0 \end{cases}.$$

ii. Two distinct solutions $(x_1, y_1) \neq (x_2, y_2) \in \mathbb{Z}^2$ of the equation $f(X, Y) = m$ such that $0 \le y_i \le U$ belong to the same $\mathcal{O}^\times_{\Delta,1}$ orbit if and only if $y_1 = y_2 = 0$ *or* $y_1 = y_2 = U$.

Corollary 4.6. There is only a finite number of $\mathcal{O}^\times_{\Delta,1}$ orbits of integral solutions of the equation $f(X, Y) = m$. There is an algorithm to list a set of representatives of the orbits.

Proof of Corollary 4.6. By Theorem 4.5i, every $\mathcal{O}^\times_{\Delta,1}$ orbit of integral solutions of the equation $f(X, Y) = m$ contains an element in the finite set $\mathcal{X} = \{(x, y) \in \mathbb{Z}^2 | f(x, y) = m \text{ and } 0 \le y \le U\}$. The elements of $\mathcal{X}$ can be listed and then sorted into orbits using Theorem 4.5ii. ∎

Proof of Theorem 4.5. Let $(u, v) \in \mathbb{Z}^2$ satisfy the equation $f(u, v) = m$. If necessary, replace (u, v) with $-(u, v)$ (which is in the same $\mathcal{O}^\times_{\Delta,1}$ orbit since $-1 \in \mathcal{O}^\times_{\Delta,1}$) so that $L = ua + v(b + \sqrt{\Delta})/2 > 0$. Let $\tau^k L = ax_k + ((b + \sqrt{\Delta})/2)y_k$, where $(x_k, y_k) \in \mathbb{Z}^2$, for all $k \in \mathbb{Z}$. By Corollary 3.16 the $\mathcal{O}^\times_{\Delta,1}$ orbit of (u, v) is precisely the set $\{\pm(x_k, y_k) | k \in \mathbb{Z}\}$. We prove next that $|y_k| \le U$ if and only if $|am/\tau|^{1/2} \le \tau^k L \le |am\tau|^{1/2}$.

Note that $L\sigma(L) = am$ and that $\sigma(\tau^k L) = \tau^{-k}\sigma(L) = ax_k + ((b - \sqrt{\Delta})/2)y_k$. It follows that $\sqrt{\Delta}\, y_k = \tau^k L - \sigma(\tau^k L) = \tau^k L - am/\tau^k L$. So we can study y_k by studying the continuous function $g(t) = t - am/t$ for $t > 0 \in \mathbb{R}$. If $am > 0$, then $g(t)$ is monotone increasing. If $am < 0$, then $g(t)$ decreases for $t < |am|^{1/2}$, increases for $t > |am|^{1/2}$, and is positive for all $t > 0$. In both cases, $|g(t)| = U\sqrt{\Delta}$ for $t = |am/\tau|^{1/2}$ and $t = |am\tau|^{1/2}$. We can conclude that $|g(t)| \le U\sqrt{\Delta}$ if and only if $|am/\tau|^{1/2} \le t \le |am\tau|^{1/2}$. We get what we want by taking $t = \tau^k L$.

Now let k be the unique integer such that $|am/\tau|^{1/2} \le \tau^k L < |am\tau|^{1/2}$. There are two cases to analyze. We will say that $(x, y) \in \mathbb{Z}^2$ is *reduced* iff $f(x, y) = m$ and $0 \le y \le U$.

Case 1. $|am/\tau|^{1/2} < \tau^k L$. In this case k is the unique integer such that $|y_k| \le U$, and in fact $|y_k| < U$. If $y_k \neq 0$, then the unique reduced element of the $\mathcal{O}_{\Delta,1}^{\times}$ orbit of (u, v) is $(x, y) = \pm(x_k, y_k)$ with the sign chosen so that $y > 0$. If $y_k = 0$, then there are exactly two reduced elements in the orbit of (u, v), namely $(x_k, 0)$ and $(-x_k, 0)$.

Case 2. $|am / \tau|^{1/2} = \tau^k L$**.** In this case $|y_k| = |y_{k+1}| = U \neq 0$ and $|y_l| > U$ for $l \neq k, k+1$. There are exactly two reduced elements of the $\mathcal{O}_{\Delta,1}^{\times}$ orbit of (u, v), namely $(x, y) = \pm(x_k, y_k)$ and $(x', y') = \pm(x_{k+1}, y_{k+1})$ with the signs chosen so that $y = y' = U > 0$.

We have proved that every $\mathcal{O}_{\Delta,1}^{\times}$ orbit of solutions contains a reduced solution, which is the assertion of Theorem 4.5i. The preceding case analysis shows that if (x, y) and (x', y') are reduced elements in the same $\mathcal{O}_{\Delta,1}^{\times}$ orbit of solutions, then $y = y' = 0$ or $y = y' = U$, which is the "only if" half of Theorem 4.5ii.

Finally, suppose that (x, y) and $(\bar{x}, y)$ are distinct integral solutions of the equation $f(X, Y) = m$. If $y = U$, then the preceding argument applied to $(u, v) = (x, y)$ leads to Case 2 and the existence of a solution $(x', y) \neq (x, y)$ in the orbit of (x, y). Since the equation $f(X, y) = m$ has at most two solutions for X, we must have $(x', y) = (\bar{x}, y)$. If $y = 0$, then $(\bar{x}, y) = -(x, y)$. In both cases (x, y) and $(\bar{x}, y)$ belong to the same $\mathcal{O}_{\Delta,1}^{\times}$ orbit of solutions. This completes the proof of Theorem 4.5ii. ■

To find a set of representatives of the $\mathcal{O}_{\Delta,1}^{\times}$ orbits of integral solutions of the equation $f(X, Y) = m$ by the method of Theorem 4.5, we must find for each integer y such that $0 \leq y \leq U$ all *integers* x such that $f(x, y) = m$. If $f(x, y) = m$, then $\Delta y^2 + 4am = (2ax + by)^2$, so $x = (x' - by)/2a$ or $x = (-x' - by)/2a$, where $x' = \sqrt{\Delta y^2 + 4am}$. A necessary but not sufficient condition for x to be integral is that $\Delta y^2 + 4am$ be a square.

As an example, we find the set $\mathcal{S}$ of all integral solutions of the equation $17X^2 + 32XY + 14Y^2 = 9$. We have $\Delta = 72$, $\tau_\Delta = \epsilon_\Delta = 17 + 4\sqrt{18}$, and $U = |(17 \cdot 9)/72 \cdot 32|^{1/2} \approx 8.246$. In the range $0 \leq y \leq 8$, $\Delta y^2 + 4am = 6^2(2y^2 + 17)$ is a square only for $y = 2$ and 4. We find that there are exactly two $\mathcal{O}_{\Delta,1}^{\times}$ orbits of solutions and that $\{(-1,2), (-5,4)\}$ is a set of representatives. The full set of integral solutions is thus $\{\pm\tau^k(-1,2), \pm\tau^k(-5,4) | k \in \mathbb{Z}\}$. The calculation (4.3) shows that $\tau \cdot (x, y) = (x', y')$, where $(x', y') = (x, y)\begin{pmatrix} -47 & 68 \\ -56 & 81 \end{pmatrix}$. The final description of $\mathcal{S}$ is as explicit as could be desired:

$$\mathcal{S} = \{\pm(-1,2)T^k, \pm(-5,4)T^k | k \in \mathbb{Z}\}, \quad \text{where } T = \begin{pmatrix} -47 & 68 \\ -56 & 81 \end{pmatrix}.$$

Exercises

1. Let f be a primitive integral binary quadratic form of nonsquare discriminant Δ. Prove that $\mathcal{O}_\Delta = \{\alpha \in \mathbb{Q}(\sqrt{\Delta}) | \alpha M_f \subset M_f\}$.

2. Verify the calculation (4.3).

3. Let $D > 0$ be a nonsquare integer and let $m \neq 0 \in \mathbb{Z}$. Let $(x, y), (x', y') \in \mathbb{Z}^2$ be two solutions of the equation $X^2 - DY^2 = m$ such that $x \equiv x'$ and $y \equiv y' \pmod{m}$. Show that (x, y) and (x', y') are in the same $\mathcal{O}_{4D,1}^{\times}$ orbit of solutions of $X^2 - DY^2 = m$. Deduce that there is only a finite number of such orbits.

4. Find all integral solutions of the equations $X^2 + 4XY - Y^2 = m$ with $-5 \leq m \leq 10$.

5. Find all integral solutions of the equations $X^2 - 3Y^2 = 13$, $X^2 - 3Y^2 = 600$, $X^2 - 82Y^2 = 2$, and $X^2 + 3XY - 5Y^2 = 65$.

6. Let $f(X, Y) = aX^2 + bXY + cY^2$ be an integral form of discriminant $\Delta = m^2$ with $m \in \mathbb{Z}$. Prove that f is the product of two linear polynomials with integer coefficients. If $a \neq 0$, show that $f(x, y) = \operatorname{sign}(a) \cdot \operatorname{GCD}(a, b, c)((a/\alpha)X + ((b + m)/2\alpha)Y)((a/\beta)X + ((b - m)/2\beta)Y)$, where $\alpha = \operatorname{GCD}(a, (b + m)/2)$ and $\beta = \operatorname{GCD}(a, (b - m)/2)$.

7. Find all integral solutions of the equation $2X^2 - XY - 3Y^2 = 8$.

8. Let $a, b, c \in \mathbb{Z}$. Show that there exists $(x, y) \neq (0, 0) \in \mathbb{Z}^2$ such that $ax^2 + bxy + cy^2 = 0$ if and only if $\Delta = b^2 - 4ac$ is a square.

9. Let f be an integral form of discriminant Δ. Prove that there exists $m \neq 0 \in \mathbb{Z}$ such that $f(X, Y) = m$ has an infinite number of integral solutions if and only if $\Delta = 0$ or Δ is a positive nonsquare.

10. Let f be an integral binary quadratic form of nonsquare discriminant $\Delta > 0$ and suppose that $N(\epsilon_\Delta) = -1$. Let $m > 0 \in \mathbb{Z}$.

i. Show that $\mathcal{O}_\Delta^{\times}$ acts on the set $\mathscr{W} = \{(x, y) \in \mathbb{Z}^2 \mid |f(x, y)| = m\}$. State and prove a variant of Theorem 4.5 that gives a set of representatives for the $\mathcal{O}_\Delta^{\times}$ orbits in $\mathscr{W}$. (Use ϵ_Δ in place of τ_Δ.)

ii. Let $\{(x_1, y_1), \ldots, (x_N, y_N)\}$ be a set of representatives of the $\mathcal{O}_\Delta^{\times}$ orbits in $\mathscr{W}$, where

$$f(x_i, y_i) = \begin{cases} m & \text{for } 1 \leq i \leq r \\ -m & \text{for } r + 1 \leq i \leq N \end{cases}.$$

Show that $\{(x_i, y_i) \mid 1 \leq i \leq r\} \cup \{\epsilon_\Delta \cdot (x_i, y_i) \mid r + 1 \leq i \leq N\}$ is a set of representatives for the $\mathcal{O}_{\Delta,1}^{\times}$ orbits of integral solutions of the equation $f(X, Y) = m$.

iii. Explain why the integral solutions of the equation $f(X, Y) = m$ can be found more efficiently by an algorithm based on i and ii than by the algorithm of the text that is based directly on Theorem 4.5.

11. Let $D \in \mathbb{Z}$ be a positive nonsquare, let $m \neq 0 \in \mathbb{Z}$, and let $\mathscr{Y} = \{(x, y) \in \mathbb{Z}^2 | x^2 - Dy^2 = m\}$. Let $\mathscr{Y}^*$ be the set of *proper* integral solutions of the equation $X^2 - DY^2 = m$, by definition $\mathscr{Y}^* = \{(x, y) \in \mathscr{Y} | \text{GCD}(x, y) = 1\}$.

  i. Show that if (x, y) and $(x', y') \in \mathscr{Y}$ are in the same $\mathscr{O}_{4D,1}^{\times}$ orbit, then $\text{GCD}(x, y) = \text{GCD}(x', y')$. Thus $\mathscr{O}_{4D,1}^{\times}$ acts on $\mathscr{Y}^*$.

  ii. Show that there is a function ψ: $\mathscr{Y}^* \to \mathbb{Z}/m$ defined by the equation $y\psi(x, y) \equiv x \pmod m$. Prove that $\psi(x, y)^2 \equiv D \pmod m$ for all $(x, y) \in \mathscr{Y}^*$. Prove that $\psi(x, y) = \psi(x' y') \in \mathbb{Z}/m$ for (x, y), $(x', y') \in \mathscr{Y}^*$ if and only if (x, y) and (x', y') belong to the same $\mathscr{O}_{4D,1}^{\times}$ orbit.

12. Let D be a positive nonsquare integer.
For $\ell, m \in \mathbb{Z}$ such that $m \neq 0$ and $\ell^2 \equiv D \pmod m$, let $X_{\ell, m} = \{(x, y) \in \mathbb{Z}^2 | x^2 - Dy^2 = m,\ \text{GCD}(x, y) = 1, \text{ and } y\ell \equiv x \pmod m\}$. This exercise shows how to reduce the computation of $X_{\ell, m}$ to the case $|\ell| \leq |m/2|$ and $|m| < \sqrt{D}$.

  i. Show that $X_{\ell, m} = X_{\ell', m}$ if $\ell \equiv \ell' \pmod m$.

  ii. Let ℓ, m, m' be such that $\ell^2 = D + mm'$. Prove that there is a bijection ϕ: $X_{\ell, m} \to X_{\ell, m'}$ given by the rule $\phi(x, y) = (x', y')$, where $(x + y\sqrt{D})(x' + y'\sqrt{D}) = \ell + \sqrt{D}$. Show that if $|\ell| \leq |m/2|$ and $|m| > \sqrt{D}$, then $|m'| < |m|$.

  iii. Using i and ii, find all integral solutions of the equations $X^2 - 17Y^2 = 757$, $X^2 - 17Y^2 = 16883$ and $X^2 - 15Y^2 = 61$. (Note. 757 is prime, $120^2 \equiv 17 \pmod{757}$. 16883 is prime, $130^2 \equiv 17 \pmod{16883}$.)

13. Write and test a computer program to perform as follows:
Input: An integral binary quadratic form f and an integer m.
Output: An explicit description of the set $\{(x, y) \in \mathbb{Z}^2 | f(x, y) = m\}$.

14. Let $(a, b) \neq (0, 0) \in \mathbb{Z}^2$ and let $m \in \mathbb{Z}$. Show that the group $\{(x, y) \in \mathbb{Z}^2 | ax + by = 0\}$, which is isomorphic to $\mathbb{Z}$, acts on the set $\{(x, y) \in \mathbb{Z}^2 | ax + by = m\}$ and that the number of orbits is 0 or 1.

5. Automorphisms

The real significance of the Pell equation is that its integral solutions give automorphisms of quadratic forms. It is the main aim of Section 5 to explain how.

Definition 5.1. A matrix $\gamma \in \mathbf{GL}_2(\mathbb{Z})$ is said to be an *automorphism* of an integral binary form $f = ax^2 + bxy + cy^2$ iff $\gamma f = f$ or equivalently, iff

$$\gamma \cdot \begin{pmatrix} a & b/2 \\ b/2 & c \end{pmatrix} \cdot {}^t\gamma = \begin{pmatrix} a & b/2 \\ b/2 & c \end{pmatrix}.$$

We say that γ is *proper* iff $\det(\gamma) = +1$ and that γ is *improper* iff $\det(\gamma) = -1$.

Let $\mathcal{A}ut(f)$ be the group of automorphisms of f.

Let $\mathcal{A}ut^+(f)$ be the group of proper automorphisms of f.

***Definition* 5.2.** Let $f = ax^2 + bxy + cy^2$ be an integral form of discriminant $\Delta \neq 0$. Define $\alpha_f(u, v) \in \mathbf{GL}_2(\mathbb{Z})$ for $(u, v) \in \mathcal{P}ell^{\pm}(\Delta)$ by the formula

$$\alpha_f(u, v) = \begin{cases} \begin{pmatrix} u - \dfrac{b}{2}v & av \\ -cv & u + \dfrac{b}{2}v \end{pmatrix} & \text{if } \Delta \equiv 0 \pmod 4 \\ \begin{pmatrix} u + \dfrac{1-b}{2}v & av \\ -cv & u + \dfrac{1+b}{2}v \end{pmatrix} & \text{if } \Delta \equiv 1 \pmod 4 \end{cases}$$

Note that $\alpha_f(u, v)$ has integer entries because $b \equiv \Delta \pmod 2$. A calculation shows that $\det(\alpha_f(u, v)) = f_\Delta(u, v)$. It is straightforward but tedious to check that α_f: $\mathcal{P}ell^{\pm}(\Delta) \to \mathbf{GL}_2(\mathbb{Z})$ is a group homomorphism. The definition of α_f should be compared with (4.3).

Theorem 5.3. Let f be an integral binary quadratic form of discriminant $\Delta \neq 0$.

i. $\alpha_f(u, v) \in \mathcal{A}ut^+(f)$ for all $(u, v) \in \mathcal{P}ell(\Delta)$.

ii. If f is a primitive form, then α_f: $\mathcal{P}ell(\Delta) \to \mathcal{A}ut^+(f)$ is a group isomorphism.

Proof. i. Straightforward calculation.

ii. Since α_f: $\mathcal{P}ell(\Delta) \to \mathcal{A}ut^+(f)$ is a homomorphism that is clearly injective, we must simply show that if f is primitive, then α_f is surjective.

Let $f = ax^2 + bxy + cy^2$ be a primitive form of discriminant $\Delta \neq 0$ and let $\gamma = \begin{pmatrix} r & s \\ t & p \end{pmatrix} \in \mathcal{A}ut^+(f)$. The equality

$$\gamma \cdot \begin{pmatrix} a & b/2 \\ b/2 & c \end{pmatrix} = \begin{pmatrix} a & b/2 \\ b/2 & c \end{pmatrix} \cdot {}^t\gamma^{-1}$$

yields the three equations

$$a(p - r) = bs, \tag{5.4a}$$

$$cs = -at, \tag{5.4b}$$

$$c(r - p) = bt. \tag{5.4c}$$

If $a = 0$, then (easy exercise) $\gamma = \pm\begin{pmatrix} 1 & 0 \\ 0 & 1 \end{pmatrix} = \alpha_f(\pm(1,0))$.

Suppose that $a \neq 0$. Then $a | bs$ and $a | cs$ by (5.4a) and (5.4b), and of course $a | as$. Since $\text{GCD}(a, b, c) = 1$ because f is primitive, we can conclude that $a | s$. Thus there exists $v \in \mathbb{Z}$ such that $s = av$. It follows from (5.4b) that $t = -cv$. If Δ is even, let $u = r + (b/2)v$; if Δ is odd let $u = r - ((1 - b)/2)v$. In both cases $u \in \mathbb{Z}$. From (5.4a) we see that $p = r + bv$, whence $p = u + (b/2)v$ if Δ is even and $p = u + ((1 + b)/2)v$ if Δ is odd. The equation $1 = \det(\gamma) = f_\Delta(u, v)$ shows that $(u, v) \in \mathscr{P}ell(\Delta)$. Thus $\gamma = \alpha_f(u, v)$. ■

Corollary 5.5. Let f be a primitive integral binary quadratic form of discriminant $\Delta \neq 0$. Then

$$\mathscr{A}ut^+(f) \simeq \begin{cases} \mathbb{Z}/2\mathbb{Z} & \text{if } \Delta \text{ is a positive square or } \Delta < -4 \\ \mathbb{Z}/4\mathbb{Z} & \text{if } \Delta = -4 \\ \mathbb{Z}/6\mathbb{Z} & \text{if } \Delta = -3 \\ \mathbb{Z}/2\mathbb{Z} \oplus \mathbb{Z} & \text{if } \Delta \text{ is a positive nonsquare} \end{cases}.$$

Proof. Theorem 5.3ii, Corollary 3.16, and Exercise 3.13. ■

Theorem 5.3ii suggests the possibility of deriving the theory of the Pell equation from a study of automorphisms of quadratic forms. This possibility will be realized in Section 9. In particular, the existence of nontrivial solutions of the Pell equation in the case of positive nonsquare discriminant Δ will follow from the existence of nontrivial proper automorphisms of the Pell form f_Δ.

Improper automorphisms are also of interest.

Theorem 5.6. Let f be an integral binary quadratic form of nonzero discriminant and let γ be an improper automorphism of f.

Then $\gamma^2 = I$ and $\gamma\beta\gamma = \beta^{-1}$ for every $\beta \in \mathscr{A}ut^+(f)$.

Proof. We prove first that $\gamma^2 = I$ for every improper automorphism γ of f. By the Cayley–Hamilton Theorem, γ satisfies its characteristic equation, which is $Z^2 - \text{tr}(\gamma)Z + \det(\gamma)I = 0$. Since $\det(\gamma) = -1$, it will suffice to prove that $\text{tr}(\gamma) = 0$.

Suppose more generally that $\gamma = \begin{pmatrix} r & s \\ t & p \end{pmatrix} \in \mathbf{GL}_2(\mathbb{C})$, that $\det(\gamma) = -1$, and that

$$\gamma \cdot \begin{pmatrix} a & b/2 \\ b/2 & c \end{pmatrix} \cdot {}^t\gamma = \begin{pmatrix} a & b/2 \\ b/2 & c \end{pmatrix}$$

for some $a, b, c \in \mathbb{C}$ with a, b, c not all zero. The 2×2 matrix equation

$$\gamma \cdot \begin{pmatrix} a & b/2 \\ b/2 & c \end{pmatrix} = \begin{pmatrix} a & b/2 \\ b/2 & c \end{pmatrix} \cdot {}^t\gamma^{-1}$$

gives four scalar equations: $a(r+p) = c(r+p) = 0$ and $br = at - cs = -bp$. It follows that $r + p = \mathrm{tr}(\gamma) = 0$.

Now let f be an integral form of nonzero discriminant, let γ be an improper automorphism of f, and let $\beta \in \mathscr{Aut}^+(f)$. Then $\gamma\beta$ is an improper automorphism of f, whence $(\gamma\beta)(\gamma\beta) = I$. It follows that $\gamma\beta\gamma = \beta^{-1}$. ■

Exercises

1. Prove that α_f is a homomorphism. Try to do this without computations beyond that of (4.3).

2. Prove Theorem 5.3i. Complete the case $a = 0$ of the proof of Theorem 5.3ii.

3. Determine $\mathscr{Aut}^+(2X^2 + 6XY + Y^2)$.

4. Let $f = ax^2 + bxy + cy^2$ be an integral form of discriminant $\Delta \neq 0$.
   i. Let $\mathscr{Aut}^*(f) = \{\gamma \in \mathbf{GL}_2(\mathbb{Z}) | \gamma f = \det(\gamma) f\}$. Show that $\mathscr{Aut}^*(f)$ is a group, that $\alpha_f(u, v) \in \mathscr{Aut}^*(f)$ for all $(u, v) \in \mathscr{Pell}^{\pm}(\Delta)$, and that if f is primitive, then α_f: $\mathscr{Pell}^{\pm}(\Delta) \to \mathscr{Aut}^*(f)$ is a group isomorphism.
   ii. Suppose that f is primitive. Prove that the following two assertions are equivalent. (a) f and $-ax^2 + bxy - cy^2$ are properly equivalent. (b) The negative Pell equation $f_\Delta(u, v) = -1$ has an integral solution $(u, v) \in \mathbb{Z}^2$.

5. Let f be an integral form of discriminant $\Delta \neq 0$, and let $\gamma \in \mathbf{SL}_2(\mathbb{Z})$. Show that $\alpha_{\gamma f}(u, v) = \gamma \cdot \alpha_f(u, v) \cdot \gamma^{-1}$ for all $(u, v) \in \mathscr{Pell}^{\pm}(\Delta)$.

6. Let $x, y, A, B, k \in \mathbb{Z}$, where $\mathrm{GCD}(x, y) = 1$ and $k \neq 0$. Prove that there exists $\ell \in \mathbb{Z}$ such that $\ell x \equiv A \pmod{k}$ and $\ell y \equiv B \pmod{k}$ if and only if $Ay \equiv Bx \pmod{k}$. Prove moreover that if such ℓ exists, then ℓ is uniquely defined mod k (i.e., prove that the two congruences have a unique solution $\bar{\ell} \in \mathbb{Z}/k$).

7. Let $f(X, Y) = aX^2 + bXY + cY^2$ be an integral form of discriminant $\Delta \neq 0$ and let $m \neq 0 \in \mathbb{Z}$.
   Let $\mathscr{P} = \mathscr{P}_{f,m} = \{(x, y) \in \mathbb{Z}^2 | f(x, y) = m$ and $\mathrm{GCD}(x, y) = 1\}$, by definition the set of *proper* solutions of the equation $f(X, Y) = m$.
   - **i.** Show that there is a well-defined map $\lambda = \lambda_{f,m}: \mathscr{P} \to \mathbb{Z}/2m$ given by the rule $\lambda(x, y) = \bar{\ell}$, where $\ell x \equiv bx + 2cy \pmod{2m}$ and $\ell y \equiv -(2ax + by) \pmod{2m}$. (See Exercise 6.)
   - **ii.** Let $(x, y) \in \mathscr{P}$ and let $\gamma = \begin{pmatrix} x & y \\ r & s \end{pmatrix} \in \mathbf{SL}_2(\mathbb{Z})$. Show that $\lambda(x, y) = \overline{B}$, where $\gamma f = mX^2 + BXY + CY^2$. Conclude that $\lambda^2 \equiv \Delta \pmod{4m}$.
   - **iii.** Let $(x, y) \in \mathscr{P}$ and let $\tau \in \mathscr{A}ut^+(f)$. Prove that $(x, y)\tau \in \mathscr{P}$ and that $\lambda((x, y)\tau) = \lambda(x, y)$. Thus the group $\mathscr{A}ut^+(f)$ acts on the right of $\mathscr{P}$ and λ is constant on the orbits. (*Suggestion*: Use ii.)
   - **iv.** Let $(x, y), (x', y') \in \mathscr{P}$, and suppose that $\lambda(x, y) = \lambda(x', y')$. Prove that there exists $\tau \in \mathscr{A}ut^+(f)$ such that $(x', y') = (x, y)\tau$. (*Suggestion*: Show that there exist $\gamma = \begin{pmatrix} x & y \\ r & s \end{pmatrix}$ and $\gamma' = \begin{pmatrix} x' & y' \\ r' & s' \end{pmatrix} \in \mathbf{SL}_2(\mathbb{Z})$ such that $\gamma f = \gamma' f$. Take $\tau = \gamma^{-1}\gamma'$.)
   - **v.** Let $\ell \in \mathbb{Z}$ be such that $\ell^2 \equiv \Delta \pmod{4m}$. Let $g = mX^2 + \ell XY + ((\ell^2 - \Delta)/4m)Y^2$. Prove that there exists $(x, y) \in \mathscr{P}$ such that $\lambda(x, y) = \bar{\ell}$ if and only if f is properly equivalent to g. Suppose that $g = \gamma f$ where $\gamma = \begin{pmatrix} r & s \\ t & u \end{pmatrix} \in \mathbf{SL}_2(\mathbb{Z})$. Show that $(r, s) \in \mathscr{P}$ and that $\lambda(r,s) = \bar{\ell}$.
   - **vi.** Suppose that f is a primitive form. Show that two proper solutions (x, y) and $(x', y') \in \mathscr{P}$ belong to the same $\mathcal{O}_{\Delta,1}^{\times}$ orbit if and only if $\lambda(x, y) = \lambda(x', y')$. Thus the $\mathcal{O}_{\Delta,1}^{\times}$ orbits in $\mathscr{P}$ are determined by congruence conditions.

6. Reduction of Indefinite Forms

Fix for this section a positive nonsquare discriminant Δ.

The problems posed in this section are practical ones. We want a finite procedure that will determine the set of proper equivalence classes of integral binary quadratic forms of discriminant Δ and we want a method to determine whether any two given integral forms of discriminant Δ are properly equivalent.

All forms considered in this section will be integral of discriminant Δ. We will write $f = [a, b, c]$ to indicate the integral form $f = aX^2 + bXY + cY^2$ of discriminant Δ. Since Δ is not a square in this section, we will always have $ac \neq 0$.

Definition 6.1. *The right neighbor* Rf of an integral form $f = [a, b, c]$ of positive nonsquare discriminant Δ is the form $[a', b', c']$ determined by the three conditions

i. $a' = c$.
ii. $b + b' \equiv 0 \pmod{2a'}$ and $\sqrt{\Delta} - |2a'| < b' < \sqrt{\Delta}$.
iii. $b'^2 - 4a'c' = \Delta$.

Note that

$$R[a, b, c] = \begin{pmatrix} 1 & 0 \\ \delta & 1 \end{pmatrix}[c, -b, a] = \begin{pmatrix} 1 & 0 \\ \delta & 1 \end{pmatrix}\begin{pmatrix} 0 & -1 \\ 1 & 0 \end{pmatrix}[a, b, c],$$

where $b + b' = 2c\delta$. Thus f is properly equivalent to its right neighbor Rf.

We construct a sequence (f_n) of forms f_n all properly equivalent to f by taking neighbors of neighbors. For $n \geq 0$ let $f_n = R^n f$ (i.e., $f_0 = f$ and $f_n = Rf_{n-1}$ for $n \geq 1$).

Definition 6.2. An integral form $f = [a, b, c]$ of positive nonsquare discriminant Δ is said to be *reduced* iff $0 < b < \sqrt{\Delta}$ and $\sqrt{\Delta} - b < |2a| < \sqrt{\Delta} + b$.

Observe that there are only finitely many reduced forms of discriminant Δ. This is because the coefficients a and b of a reduced form $[a, b, c]$ of discriminant Δ lie within bounded intervals and c is determined by a, b, and Δ.

The basic facts of Gauss's reduction theory are as follows.

Theorem 6.3. Let Δ be a positive nonsquare discriminant.

i. Let f be an integral form of discriminant Δ. There is an integer $m \geq 0$ such that $f_m = [a, b, c]$ is a reduced form with $|a| < \frac{1}{2}\sqrt{\Delta}$. Hence every integral form of discriminant Δ is properly equivalent to a reduced form.

ii. Let f be an integral form of discriminant Δ. Then the sequence $f_0, f_1, f_2, \ldots$ contains only a finite number of distinct forms f_n.

iii. Two integral forms f and g of discriminant Δ are properly equivalent if and only if there exist nonnegative integers M and N such that $f_M = g_N$.

Corollary 6.4. There is only a finite number of proper equivalence classes of integral forms of discriminant Δ.

Proof of Corollary 6.4. By Theorem 6.3i each proper equivalence class of integral forms of discriminant Δ must contain at least one reduced form. We have already remarked that there are only finitely many such reduced forms. ■

Theorem 6.3 reduces the test for proper equivalence of two forms f and g to a finite sequence of tests for equality of f_m and g_n. One can compute the two sequences $f_0, f_1, \ldots$ and $g_0, g_1, \ldots$ until they repeat an element, which they both will eventually do by Theorem 6.3ii. By Theorem 6.3iii, the two resulting finite sequences will have an element in common if and only if f and g are properly equivalent.

For example, the forms $2X^2 + 8XY + 3Y^2$ and $X^2 - 10Y^2$ of discriminant $\Delta = 40$ are shown to be not properly equivalent by calculation of the following two sequences of right neighbors:

i. $[2,8,3]$, $[3,4,-2]$, $[-2,4,3]$, $[3,2,-3]$, $[-3,4,2]$, $[2,4,-3]$, $[-3,2,3]$, $[3,4,-2]$.

ii. $[1,0,-10]$, $[-10,0,1]$, $[1,6,-1]$, $[-1,6,1]$, $[1,6,-1]$.

We can use the procedure just described to partition the finite set of reduced forms of discriminant Δ into proper equivalence classes. By Theorem 6.3i we will find in this way all proper equivalence classes of integral forms of discriminant Δ.

The partition by proper equivalence of the set of reduced forms of discriminant Δ is clarified by the following supplement to Theorem 6.3.

Theorem 6.5. Let Δ be a positive nonsquare discriminant.

i. R induces a permutation of the finite set of reduced integral forms of discriminant Δ.

ii. An integral form f of discriminant Δ is reduced if and only if there exists an integer $N > 0$ such that $f_N = f$.

iii. Two reduced integral forms f and g of discriminant Δ are properly equivalent if and only if there is an integer $M \geq 0$ such that $f_M = g$.

From Theorem 6.5ii we learn that the reduced forms are precisely those f for which the sequence (f_n) is periodic, i.e., for which there is an integer $N > 0$ such that $f_{n+N} = f_n$ for all $n \geq 0$. Let f be a reduced form, and let N be minimal with the preceding property. By Theorem 6.5iii the finite sequence $f_0, f_1, \ldots, f_{N-1}$, classically called the *period* of f, contains all the reduced forms that are properly equivalent to f. Thus two reduced forms of discriminant Δ are properly equivalent if and only if their periods are the same, up to a cyclic permutation.

The permutation induced by R on the finite set of reduced forms of discriminant Δ can be expressed canonically as a product of disjoint cycles. It is easy to do this explicitly for any given Δ. The disjoint cycles, which are nothing but the periods of the reduced forms, are in one-to-one correspondence with the proper equivalence classes of forms of discriminant Δ.

The result is really quite pretty. We begin with an action of the unwieldy infinite group $\mathbf{SL}_2(\mathbb{Z})$ on the infinite set of integral forms of discriminant Δ. We introduce the action of a finite cyclic group (generated by R) on a finite set (the subset of reduced forms). Then Theorems 6.3 and 6.5 tell us that the orbits for the two group actions are the same.

We turn to the proofs of Theorems 6.3 and 6.5. The proof of Theorem 6.5iii, which is based on continued fraction considerations, will be deferred to Section 8. The proof of Theorem 6.3iii given in this section depends upon Theorem 6.5iii.

We first note a symmetry in the definition of a reduced form.

Lemma 6.6. For an integral form $[a, b, c]$ of positive nonsquare discriminant Δ the following three assertions are equivalent.

1. $[a, b, c]$ is reduced.
2. $0 < b < \sqrt{\Delta}$ and $\sqrt{\Delta} - b < |2c| < \sqrt{\Delta} + b$.
3. $[c, b, a]$ is reduced.

Proof. The lemma follows easily from the calculation $|2a| \cdot |2c| = (\sqrt{\Delta} - b)(\sqrt{\Delta} + b)$, which holds for $[a, b, c]$ if $0 < b < \sqrt{\Delta}$. ■

Proof of Theorem 6.3i. Let f be an integral form of discriminant Δ and let $f_n = [a_n, b_n, c_n]$ for $n \geq 0$. There must exist an integer $m \geq 1$ such that $|a_m| \leq |a_{m+1}|$, because an infinite descending sequence of positive integers $|a_1| > |a_2| > |a_3| > \cdots$ is impossible. We claim that for such m, $f_m = [a_m, b_m, a_{m+1}]$ is reduced and that $|a_m| < \frac{1}{2}\sqrt{\Delta}$.

Indeed, let $f = [a, b, c]$ be an integral form of discriminant Δ such that $\sqrt{\Delta} - |2a| < b < \sqrt{\Delta}$ and $|a| \leq |c|$. Then $0 < \sqrt{\Delta} - b < |2a| \leq |2c| = |(\Delta - b^2)/2a| < |\sqrt{\Delta} + b|$, from which it follows that $b > 0$. Thus $b^2 < \Delta$ and so $|2a|^2 \leq |4ac| = \Delta - b^2 < \Delta$. Hence $|2a| < \sqrt{\Delta} < \sqrt{\Delta} + b$. Therefore the form $[a, b, c]$ is reduced and $|a| < \frac{1}{2}\sqrt{\Delta}$. ■

Proof of Theorem 6.5i. We first show that the right neighbor of a reduced form is reduced.

Let $f = [a, b, c]$ be a reduced form of discriminant Δ and let $Rf = [a', b', c']$. To prove that Rf is reduced we must show that $b' > 0$ and $|2a'| < \sqrt{\Delta} + b'$.

Let $k \in \mathbb{Z}$ be defined by the equation $b + b' = k|2a'|$. We can write $b + b' = ((\sqrt{\Delta} + b) - |2c|) + (b' - (\sqrt{\Delta} - |2a'|))$, where the two terms on the right are positive by Lemma 6.6 and Definition 6.1ii. This shows that $k \geq 1$. It follows as desired that $2b' = (\sqrt{\Delta} - b) + (b' - (\sqrt{\Delta} - |2a'|)) + (k - 1)|2a'|$

> 0 and that $\sqrt{\Delta} + b' - |2a'| = (\sqrt{\Delta} - b) + (k-1)|2a'| > 0$. Thus Rf is reduced.

Now let X be the set of reduced integral forms of discriminant Δ. We will prove that R: $X \to X$ is a bijection by finding an inverse for R.

Define the *left neighbor* Lf of an integral form f of discriminant Δ to be the form $Lf = \begin{pmatrix} 0 & 1 \\ 1 & 0 \end{pmatrix} R \begin{pmatrix} 0 & 1 \\ 1 & 0 \end{pmatrix} f$. By Lemma 6.6 and the fact that R maps reduced forms to reduced forms, we see that Lf is reduced if f is reduced.

Now compute the composition RL. First compute

$$L[a, b, c] = \begin{pmatrix} 0 & 1 \\ 1 & 0 \end{pmatrix} R[c, b, a] = \begin{pmatrix} 0 & 1 \\ 1 & 0 \end{pmatrix} [a, B, *] = [*, B, a],$$

where $B \equiv -b \pmod{2a}$, $\sqrt{\Delta} - |2a| < B < \sqrt{\Delta}$, and $*$ is determined by B, a, and Δ. Thus $RL[a, b, c] = [a, \beta, *]$, where $\beta \equiv -B \equiv b \pmod{2a}$ and $\sqrt{\Delta} - |2a| < \beta < \sqrt{\Delta}$. If $f = [a, b, c]$ is reduced, then also $\sqrt{\Delta} - |2a| < b < \sqrt{\Delta}$, from which it follows that $\beta = b$ and hence that $R(Lf) = f$. We have proved that the map R: $X \to X$ is surjective. Since X is finite, R is bijective and L is its inverse. ■

Proof of Theorem 6.3ii. Let f be an integral form of discriminant Δ. By Theorem 6.3i there is an m such that f_m is reduced. By Theorem 6.5i we conclude that f_n is reduced for every $n \geq m$. Since there is only a finite number of reduced forms of discriminant Δ, there can be only a finite number of distinct f_n. ■

Proof of Theorem 6.5ii. The group of permutations of the finite set X of reduced integral forms of discriminant Δ is a finite group. Hence all its elements have finite order. Let N be the order of the permutation of X induced by R. Then $R^N f = f$ for all reduced forms f. In other notation, $f_N = f$.

Conversely, suppose that $f_N = f$ for an integral form f and positive integer N. Then $f_{kN} = f$ for all integers $k > 0$. By Theorems 6.3i and 6.5i, the form f_{kN} is reduced for all sufficiently large k. Hence f is reduced. ■

Proof of Theorem 6.3iii. Since f is properly equivalent to f_M and g is properly equivalent to g_N, the equality $f_M = g_N$ implies that f and g are properly equivalent.

Conversely, suppose that f and g are two integral forms of discriminant Δ that are properly equivalent. By Theorem 6.3i there exist nonnegative integers m and n such that $R^m f$ and $R^n g$ are reduced. Since $R^m f$ and $R^n g$ are properly equivalent, Theorem 6.5iii (which will be proved in Section 8) implies that there exists $k \geq 0$ such that $R^k R^m f = R^n g$. Thus $f_{k+m} = g_n$. The proof is complete. ■

Exercises

1. Let $f = X^2 - 10Y^2$ and let $g = -201X^2 + 244XY - 74Y^2$. Show that f and g are properly equivalent and find $\gamma \in \mathbf{SL}_2(\mathbb{Z})$ such that $\gamma g = f$.

2. List all reduced forms of discriminant 65. Divide them into proper equivalence classes.

3. List all reduced forms of discriminant 85. Divide them into proper equivalence classes.

4. **i.** Show that all integral forms of discriminant 13 are properly equivalent.

ii. Prove that the form $X^2 + XY - 3Y^2$ represents a prime number p if and only if $p = 13$ or $p \equiv \pm 1, \pm 3$, or $\pm 4 \pmod{13}$.

5. Write two computer programs to perform as follows.

i. Input: Two integral forms f and f' of positive nonsquare discriminants.
Task: To determine whether f and f' are properly equivalent and if so to produce $\gamma \in \mathbf{SL}_2(\mathbb{Z})$ such that $\gamma f = f'$.

ii. Input: A positive nonsquare discriminant Δ.
Task: To produce a list containing exactly one reduced form in each proper equivalence class of integral forms of discriminant Δ.

6. Find an upper bound (depending on Δ) for the number of proper equivalence classes of integral forms of positive nonsquare discriminant Δ.

7. Prove directly that $L(Rf) = f$ for every reduced integral form f of positive nonsquare discriminant.

8. **i.** Given an integral form $f = [a, b, c]$ of positive nonsquare discriminant, let $f^* = [-c, b, -a]$ and let $f' = [-a, b, -c]$. Let $R^n f = [a_n, b_n, c_n]$ for all $n \geq 0$. Show that $L^n(f^*) = [-c_n, b_n, -a_n]$ for all $n \geq 0$. Prove that if f is reduced and $R^{2m} f = f^*$, then $R^m f = [A, B, -A]$ for some $A, B \in \mathbb{Z}$.

ii. Let p be a prime, $p \equiv 1 \pmod 4$. Let $t = [\sqrt{p}]$ and let $f = [1, 2t, t^2 - p]$. Show that f is a reduced form of discriminant $4p$ and that f is properly equivalent to f' (Exercises 3.8iii and 5.4ii). Show that $Lf' = f^*$. Conclude that there exists an integer $m \geq 0$ such that $R^m f = [a, b, -a]$ for some $a, b \in \mathbb{Z}$. Since $p = a^2 + (b/2)^2$, we have a new proof of Fermat's theorem that every prime congruent to 1 mod 4 is a sum of two squares.

Express $p = 233$ as a sum of two squares by computing the sequence $f_0, f_1, f_2, \ldots$.

iii. Let p be a prime, $p \equiv 1 \pmod 4$. Let $s = [(\sqrt{p} - 1)/2]$ and let $f = [1, 2s + 1, s^2 + s + (1 - p)/4]$. Show that f is a reduced form of discriminant p, that f is properly equivalent to f', and that $Lf' = f^*$. Prove that there exists an integer $m \geq 0$ such that $R^m f = [a, b, -a]$ for some $a, b \in \mathbb{Z}$ and that $p = (2a)^2 + b^2$.
Express $p = 233$ as a sum of two squares by computing the sequence $f_0, f_1, f_2, \ldots$. Do the same for $p = 2837$.

9. Let f be an integral form of positive nonsquare discriminant Δ and let $m \neq 0 \in \mathbb{Z}$ be such that $|m| < \frac{1}{2}\sqrt{\Delta}$. Let $\mathscr{P}_{f,m}$ be the set of proper integral solutions of the equation $f(X, Y) = m$ (as in Exercise 5.7, which the present exercise continues).

i. Prove that $\mathscr{P}_{f,m} \neq \varnothing$ if and only if there exists $n \geq 0$ such that $R^n f = [m, b, c]$ for some $b, c \in \mathbb{Z}$.

ii. Let $\ell \in \mathbb{Z}$. Prove that there exists $(x, y) \in \mathscr{P}_{f,m}$ such that $\lambda(x, y) = \bar{\ell}$ if and only if there exists $n \geq 0$ such that $R^n f = [m, b, c]$ for some $b, c \in \mathbb{Z}$ with $b \equiv \ell \pmod{2m}$.

iii. Produce a fast algorithm based on i and ii to find a set of representatives of the $\mathscr{A}ut^+(f)$ orbits in $\mathscr{P}_{f,m}$ (for the case $|m| < \frac{1}{2}\sqrt{\Delta}$).

iv. Find two ways all integral solutions of the equation $3X^2 + 2XY - 4Y^2 = 3$, first by the method of iii, and second by the method of Theorem 4.5.

10. **i.** Give an algorithm based on Exercise 5.7v and Theorem 6.3 to perform as follows.
Input: An integral form f of positive nonsquare discriminant and an integer m.
Output: A set of representatives of the $\mathscr{A}ut^+(f)$ orbits of proper integral solutions of the equation $f(X, Y) = m$.
(The first step of the algorithm will be to find all solutions of the congruence $\lambda^2 \equiv \Delta \pmod{4m}$.)

ii. Use the preceding algorithm to find all integral solutions (proper or not) of the equation $17X^2 + 32XY + 14Y^2 = 9$.

11. Let $f = [a, b, c]$ be an integral form of positive nonsquare discriminant and let $\bar{f} = [c, b, a]$. Prove that the following three assertions are equivalent. (a) f has an improper automorphism. (b) f and $\bar{f}$ are properly equivalent. (c) f is properly equivalent to a form $[A, B, C]$ such that $A|B$. (*Suggestion*: Show that if $f = [a, b, c]$ is reduced and $R^{2i+1}[a, b, c] = [c, b, a]$, then $[A, B, C] = R^{i+1}[a, b, c]$ has property (c).)

12. Let $f = [a, b, c]$ be an integral form of positive nonsquare discriminant Δ and let $Rf = [a', b', c']$. Show that $b' = 2c\delta - b$ and $c' = \delta(b' - b)/2 + a$, where $\delta = \operatorname{sign}(c)[(b + \sqrt{\Delta})/|2c|] = \operatorname{sign}(c)[(b + [\sqrt{\Delta}])/|2c|]$.

7. Continued Fractions

We present the elementary facts of continued fractions that we will need in Section 8 for the proof of Theorem 6.5iii.

Definition 7.1. Given real numbers $a_0, a_1, \ldots, a_n$ such that $a_i > 0$ for all $i > 0$, we define the *continued fraction* $(a_0, a_1, \ldots, a_n) \in \mathbb{R}$ recursively:

$$(a_0) = a_0 \quad \text{and} \quad (a_0, a_1, \ldots, a_n) = a_0 + \frac{1}{(a_1, a_2, \ldots, a_n)}.$$

The terminology comes from the expression

$$(a_0, a_1, \ldots, a_n) = a_0 + \cfrac{1}{a_1 + \cfrac{1}{\ddots + \cfrac{1}{a_n}}}.$$

We will say that a continued fraction $(a_0, a_1, \ldots, a_{n-1}, y)$ is *nearly simple* iff a_i is a positive integer for $0 \le i \le n-1$ and $y > 1$.

Lemma 7.2. Let $x = (a_0, a_1, \ldots, a_{n-1}, y)$ be a continued fraction. Let

$$\begin{pmatrix} r & s \\ t & u \end{pmatrix} = \begin{pmatrix} a_0 & 1 \\ 1 & 0 \end{pmatrix}\begin{pmatrix} a_1 & 1 \\ 1 & 0 \end{pmatrix} \cdots \begin{pmatrix} a_{n-1} & 1 \\ 1 & 0 \end{pmatrix} \in \mathbf{GL}_2(\mathbb{R}).$$

Then $x = (ry + s)/(ty + u)$.

Proof. By induction on $n \ge 1$. Details left as an exercise. ■

We will occasionally write $\gamma y = (ry + s)/(ty + u)$, where $\gamma = \begin{pmatrix} r & s \\ t & u \end{pmatrix} \in \mathbf{GL}_2(\mathbb{R})$ and $y \in \mathbb{R}$ are such that $ty + u \neq 0$. The inductive step in the proof of Lemma 7.2 can be seen to follow from the calculation $\gamma_1(\gamma_2 y) = (\gamma_1\gamma_2)y$ for $\gamma_1, \gamma_2 \in \mathbf{GL}_2(\mathbb{R})$ and $y \in \mathbb{R}$. It follows that $x = \gamma y$ implies $y = \gamma^{-1}x$, another formula we will have occasion to use.

For a nearly simple continued fraction, the matrix $\begin{pmatrix} r & s \\ t & u \end{pmatrix}$ of Lemma 7.2 belongs to $\mathbf{GL}_2(\mathbb{Z})$. It is through the group $\mathbf{GL}_2(\mathbb{Z})$ that nearly simple continued fractions will be linked to equivalence of binary quadratic forms.

We will need two facts from the general theory of continued fractions, an existence theorem and a uniqueness theorem. We begin with uniqueness.

Proposition 7.3. Let $x = (a_0, a_1, \ldots, a_n)$ and $y = (b_0, b_1, \ldots, b_n)$ be two nearly simple continued fractions.

If $x = y$, then $a_i = b_i$ for all i.

Proof. By induction on n, beginning with the trivial case $n = 0$.

Let n be a positive integer and let x and y be as in the statement of the lemma. The equation $x = a_0 + 1/(a_1, \ldots, a_n)$ shows that $a_0 = [x]$, for $(a_1, \ldots, a_n) > 1$ because x is nearly simple. Similarly, $b_0 = [y]$. If $x = y$, then $a_0 = b_0$. It follows that $(a_1, \ldots, a_n) = (b_1, \ldots, b_n)$ and therefore, by the inductive hypothesis, that $a_i = b_i$ for all $i > 0$. ■

Our next result is the key existence theorem.

Proposition 7.4. Let $\Gamma = \begin{pmatrix} R & S \\ T & U \end{pmatrix} \in \mathbf{GL}_2(\mathbb{Z})$ be such that $R \geq S \geq U \geq 0$ and $R \geq T \geq U$.

Then there exist positive integers $n, a_0, a_1, \ldots, a_{n-1}$ such that

$$\Gamma = \begin{pmatrix} a_0 & 1 \\ 1 & 0 \end{pmatrix} \begin{pmatrix} a_1 & 1 \\ 1 & 0 \end{pmatrix} \cdots \begin{pmatrix} a_{n-1} & 1 \\ 1 & 0 \end{pmatrix}.$$

Proof. The proof is by induction on T.

If $T = 1$, there are two cases depending on the value of U. Either

$$\Gamma = \begin{pmatrix} R & 1 \\ 1 & 0 \end{pmatrix} \quad \text{or} \quad \Gamma = \begin{pmatrix} S+1 & S \\ 1 & 1 \end{pmatrix} = \begin{pmatrix} S & 1 \\ 1 & 0 \end{pmatrix} \begin{pmatrix} 1 & 1 \\ 1 & 0 \end{pmatrix}.$$

Now suppose that $T > 1$. Since $\det(\Gamma) = \pm 1$, we must have in this case that $R > T > U > 0$ and $R/T \notin \mathbb{Z}$. Let m be the (positive) integer such that $R/T - 1 < m < R/T$ and let

$$\gamma = \begin{pmatrix} m & 1 \\ 1 & 0 \end{pmatrix}^{-1} \Gamma = \begin{pmatrix} T & U \\ R - mT & S - mU \end{pmatrix} = \begin{pmatrix} r & s \\ t & u \end{pmatrix}.$$

To complete the proof of the lemma we show that $\gamma \in \mathbf{GL}_2(\mathbb{Z})$ is a matrix to which the induction hypothesis may be applied.

It is automatic that $r > s > 0$, and the choice of m insures that $r = T > t > 0$. That $u \geq 0$ can be seen from the calculation $u = (st + \det(\gamma))/r$. It remains only to show that $s \geq u$ and that $t \geq u$. But $u > s$ would imply that $ru \geq (t+1)(s+1) = st + s + t + 1$, which would contradict the fact that $ru - st = \pm 1$. A symmetric argument rules out the possibility that $u > t$.

Hence γ can be expressed as a product of matrices of the required form. It follows that $\Gamma = \begin{pmatrix} m & 1 \\ 1 & 0 \end{pmatrix} \gamma$ can be so expressed as well. ■

We conclude Section 7 with a more specialized lemma that is designed specifically for the proof of Theorem 6.5iii. The conclusion of Lemma 7.5 should be compared with the hypothesis of Proposition 7.4.

Lemma 7.5. Let $\gamma = \begin{pmatrix} a & b \\ c & d \end{pmatrix} \in \mathbf{GL}_2(\mathbb{Z})$. Let $x, y \in \mathbb{R}$. Suppose that $cx + d \neq 0$ and $cy + d \neq 0$ and that the following two systems of inequalities hold:

$$x > 1 \quad \text{and} \quad \frac{ax + b}{cx + d} > 1, \tag{$*$}$$

$$-1 < y < 0 \quad \text{and} \quad -1 < \frac{ay + b}{cy + d} < 0. \tag{$**$}$$

Then exactly one of the following three statements is true:

A. $\gamma = \pm\begin{pmatrix} 1 & 0 \\ 0 & 1 \end{pmatrix}$.

B. $\gamma = \pm\begin{pmatrix} r & s \\ t & u \end{pmatrix}$, where $r \geq s \geq u \geq 0$ and $r \geq t \geq u$.

C. $\gamma^{-1} = \pm\begin{pmatrix} r & s \\ t & u \end{pmatrix}$, where $r \geq s \geq u \geq 0$ and $r \geq t \geq u$.

Proof. The lemma is proved by analysis of five cases.

Case 1. Suppose that $abcd \neq 0$, $c > 0$ and $d > 0$.

The nonzero integers ad and bc are consecutive, since $\det \gamma = \pm 1$, and therefore they have the same sign. Hence a and b have the same sign. Since by $(*)$ we have $ax + b > 0$, we can conclude that $a > 0$ and $b > 0$.

By $(*)$ we have $ax + b > cx + d$. Hence $(a - c)x > (d - b)$. This last inequality is incompatible with the supposition that both $a \leq c$ and $b \leq d$. Thus either $a > c$ or $b > d$. If $a > c$, we can prove that $b \geq d$ by contradiction as follows: $d > b$ implies that $ad \geq (c + 1)(b + 1) = bc + b + c + 1$, whence $\det \gamma = ad - bc > 1$, which is false. Similarly, if $b > d$, then $a \geq c$. Therefore, $a \geq c$ and $b \geq d$.

If $ay + b < 0$, then $b/a < |y| < 1$, and thus $a > b$. It follows by an argument similar to one in the previous paragraph that $c \geq d$. If $ay + b > 0$, then by $(**)$ we have $cy + d < 0$, whence $d/c < |y| < 1$, and thus $c > d$. It follows that $a \geq b$. Therefore, $a \geq b$ and $c \geq d$.

We have verified that statement B holds.

Case 2. Suppose that $abcd \neq 0$, $c < 0$ and $d < 0$. Statement B holds because $-\gamma$ belongs to Case 1.

Case 3. Suppose that $abcd \neq 0$, $c < 0$ and $d > 0$. It follows from the equation $ad - bc = \pm 1$ that a and b have opposite signs. Since $cy + d > 0$, $(**)$ implies that $ay + b < 0$. Therefore, $a > 0$.

Now consider $\gamma^{-1} = \pm\begin{pmatrix} d & -b \\ -c & a \end{pmatrix} \in \mathbf{GL}_2(\mathbb{Z})$. Let $x' = \gamma x$ and $y' = \gamma y$. We get two systems of inequalities like $(*)$ and $(**)$:

$$x' > 1 \quad \text{and} \quad \gamma^{-1}x' > 1, \tag{$*'$}$$

$$-1 < y' < 0 \quad \text{and} \quad -1 < \gamma^{-1}y' < 0. \tag{$**''$}$$

The reasoning of Case 1 or 2 is seen to apply to the matrix γ^{-1}. We conclude that statement C of the lemma holds.

Case 4. Suppose that $abcd \neq 0$, $c > 0$ and $d < 0$. Statement C holds because $-\gamma$ belongs to Case 3.

Case 5. Suppose that $abcd = 0$. A simple analysis, which is left as an exercise, shows that

$$\gamma = \pm\begin{pmatrix} 1 & 0 \\ 0 & 1 \end{pmatrix}, \qquad \gamma = \pm\begin{pmatrix} m & 1 \\ 1 & 0 \end{pmatrix},$$

or

$$\gamma = \pm\begin{pmatrix} 0 & 1 \\ 1 & -m \end{pmatrix} = \pm\begin{pmatrix} m & 1 \\ 1 & 0 \end{pmatrix}^{-1}$$

for some positive integer m. The three possibilities correspond to the three statements A, B, and C.

It remains only to prove that the three properties A, B, and C are mutually exclusive. But that is easy and may be left to the reader. ∎

Exercises

1. Prove Lemma 7.2.
2. Prove that $\gamma_1(\gamma_2 y) = (\gamma_1\gamma_2)y$ for all $\gamma_1, \gamma_2 \in \mathbf{GL}_2(\mathbb{R})$ and $y \in \mathbb{R}$ for which the expressions are defined.
3. Let $n, a_0, a_1, \ldots, a_{n-1}$ be positive integers and let

$$\Gamma = \begin{pmatrix} R & S \\ T & U \end{pmatrix} = \begin{pmatrix} a_0 & 1 \\ 1 & 0 \end{pmatrix}\begin{pmatrix} a_1 & 1 \\ 1 & 0 \end{pmatrix} \cdots \begin{pmatrix} a_{n-1} & 1 \\ 1 & 0 \end{pmatrix}.$$

   i. Prove that $R/T = (a_0, a_1, \ldots, a_{n-1})$ and that if $n \geq 2$, then $S/U = (a_0, a_1, \ldots, a_{n-2})$. By considering the transpose of Γ, find continued fraction expressions for R/S and T/U.

ii. (Converse to Proposition 7.4.) Prove that $R \geq S \geq U \geq 0$ and $R \geq T \geq U$.

iii. Let $m, b_0, b_1, \ldots, b_{m-1}$ be positive integers such that

$$\Gamma = \begin{pmatrix} b_0 & 1 \\ 1 & 0 \end{pmatrix}\begin{pmatrix} b_1 & 1 \\ 1 & 0 \end{pmatrix}\cdots\begin{pmatrix} b_{m-1} & 1 \\ 1 & 0 \end{pmatrix}.$$

Prove that $n = m$ and that $a_i = b_i$ for all i.

4. Complete the proof of Lemma 7.5 by carrying out the analysis of Case 5, the case $abcd = 0$.

5. Read the sections on continued fractions in any text that discusses the subject for its own sake.

8. Reduction (II)

Fix for this section a positive nonsquare discriminant Δ.

We are going to prove Theorem 6.5iii. We begin with a key definition.

Definition 8.1. Let $f = [a, b, c]$ be an integral form of positive nonsquare discriminant Δ. We define the *roots* $\Omega(f)$ and $\omega(f)$ of f by the formulas

$$\Omega = \frac{b + \sqrt{\Delta}}{2c} \quad \text{and} \quad \omega = \sigma(\Omega) = \frac{b - \sqrt{\Delta}}{2c}.$$

Observe that since $\sqrt{\Delta}$ is irrational, f can be recovered from knowledge of $\Omega(f)$ and Δ. In other words, if f and g are integral forms of discriminant Δ such that $\Omega(f) = \Omega(g)$, then $f = g$. In some situations we will find it more convenient to work with the single real number Ω than to work directly with the form f.

Properties of f can often be simply expressed in terms of the roots of f. We have for instance:

Lemma 8.2. f is reduced if and only if $\omega(f)\Omega(f) < 0$ and $|\omega(f)| < 1 < |\Omega(f)|$.

Proof. Let $f = [a, b, c]$. The stated inequalities on $\omega(f)$ and $\Omega(f)$ when expressed in terms of a, b, and c are $|b| < \sqrt{\Delta}$ and $|\sqrt{\Delta} - b| < |2c| < |\sqrt{\Delta} + b|$. Lemma 8.2 is thus a consequence of Lemma 6.6. ∎

The next lemma shows the way in which Ω and ω change under proper equivalence of forms.

Lemma 8.3. Let f be an integral form of positive nonsquare discriminant Δ and let $\gamma = \begin{pmatrix} r & s \\ t & u \end{pmatrix} \in \mathbf{SL}_2(\mathbb{Z})$. Then

$$\Omega(\gamma f) = \frac{r\Omega(f) + s}{t\Omega(f) + u} \quad \text{and} \quad \omega(\gamma f) = \frac{r\omega(f) + s}{t\omega(f) + u}.$$

Proof. Exercise. ∎

The roots of the forms in the sequence $R^n f$ are related by continued fractions.

Lemma 8.4. Let f and g be reduced integral forms of positive nonsquare discriminant Δ and let n be a positive integer.

The following two assertions are equivalent.

1. $R^n f = g$.
2. $\operatorname{sign}(\Omega(g)) = (-1)^n \operatorname{sign}(\Omega(f))$ and there is a nearly simple continued fraction $|\Omega(f)| = (d_0, d_1, \ldots, d_{n-1}, |\Omega(g)|)$.

Proof. $1 \Rightarrow 2$. The proof will be by induction on n. Let $f = [a, b, c]$ be a reduced integral form of discriminant Δ and let $\Omega = \Omega(f)$. Let $Rf = [a', b', c']$ and let $\Omega' = \Omega(Rf)$. We know that $cc' = a'c'$ is negative because Rf is reduced (by Theorem 6.5i). Thus $-1 = \operatorname{sign}(cc') = \operatorname{sign}(\Omega\Omega')$.

Let $k \in \mathbb{Z}$ be such that $b + b' = k|2c|$. We have the equation

$$\Omega = \frac{b + \sqrt{\Delta}}{2c} = \frac{k|2c| - b' + \sqrt{\Delta}}{2c} = k \operatorname{sign}(c) - \frac{1}{\Omega'}.$$

There are two cases. If $c > 0$, then $\Omega = |\Omega|$ and $\Omega' = -|\Omega'|$. If $c < 0$, then $\Omega = -|\Omega|$ and $\Omega' = |\Omega'|$. In both cases, we find that $|\Omega| = k + 1/|\Omega'| = (k, |\Omega'|)$. Note finally that $k \geq 1$, since b and b' are positive, and that $|\Omega'| > 1$ by Lemma 8.2. The case $n = 1$ of the implication $1 \Rightarrow 2$ is proved.

Now suppose that $n > 1$. Application of an induction hypothesis yields

$$\begin{aligned} |\Omega| &= (k, |\Omega(Rf)|) = \big(k, (k_1, k_2, \ldots, k_{n-1}, |\Omega(R^n f)|)\big) \\ &= \big(k, k_1, \ldots, k_{n-1}, |\Omega(R^n f)|\big), \end{aligned}$$

where k and all k_i are positive integers and

$$\operatorname{sign}(\Omega(R^n f)) = -\operatorname{sign}\big(\Omega(R^{n-1} f)\big) = (-1)^n \operatorname{sign}(\Omega(f)).$$

$2 \Rightarrow 1$. Let f and g be as in the statement of the theorem and suppose that 2 holds. The implication $1 \Rightarrow 2$ just proved asserts the existence of a nearly simple continued fraction $|\Omega(f)| = (k_0, k_1, \ldots, k_{n-1}, |\Omega_n|)$, where $\Omega_n = \Omega(R^n f)$. By the uniqueness assertion Proposition 7.3 we can conclude that

$|\Omega(g)| = |\Omega_n|$. But Ω_n and $\Omega(g)$ have the same sign, because $\text{sign}(\Omega_0\Omega_n) = (-1)^n$ by the implication $1 \Rightarrow 2$, and we have hypothesised that $\text{sign}(\Omega_0\Omega(g)) = (-1)^n$. Hence $\Omega_n = \Omega(g)$. Therefore, $R^n f = g$, which is what we wanted to prove. ■

At last we are ready.

Proof of Theorem 6.5iii. Let f and g be properly equivalent reduced forms of discriminant Δ and let $\tau = \begin{pmatrix} r & s \\ t & u \end{pmatrix} \in \mathbf{SL}_2(\mathbb{Z})$ be such that $f = \tau g$.

Let Ω and ω denote the roots of f and let Ω' and ω' denote the roots of g. Define $\gamma \in \mathbf{GL}_2(\mathbb{Z})$ such that $\det(\gamma) = \text{sign}(\Omega\Omega')$:

$$\gamma = \begin{cases} \begin{pmatrix} r & s \\ t & u \end{pmatrix} = \tau & \text{if } \Omega > 0, \Omega' > 0 \\ \begin{pmatrix} r & -s \\ -t & u \end{pmatrix} = \begin{pmatrix} -1 & 0 \\ 0 & 1 \end{pmatrix} \tau \begin{pmatrix} -1 & 0 \\ 0 & 1 \end{pmatrix} & \text{if } \Omega < 0, \Omega' < 0 \\ \begin{pmatrix} -r & s \\ -t & u \end{pmatrix} = \tau \begin{pmatrix} -1 & 0 \\ 0 & 1 \end{pmatrix} & \text{if } \Omega > 0, \Omega' < 0 \\ \begin{pmatrix} -r & -s \\ t & u \end{pmatrix} = \begin{pmatrix} -1 & 0 \\ 0 & 1 \end{pmatrix} \tau & \text{if } \Omega < 0, \Omega' > 0 \end{cases}. \tag{8.5}$$

Define a, b, c, d by the equation $\gamma = \begin{pmatrix} a & b \\ c & d \end{pmatrix}$. Starting with Lemma 8.3, it is easy to check that

$$|\Omega| = \frac{a|\Omega'| + b}{c|\Omega'| + d} \quad \text{and} \quad -|\omega| = \frac{a(-|\omega'|) + b}{c(-|\omega'|) + d}. \tag{8.6}$$

Lemma 8.2 now shows that γ satisfies the hypotheses of Lemma 7.5 with $x = |\Omega'|$ and $y = -|\omega'|$. Hence γ satisfies Lemma 7.5A, B, or C.

If γ satisfies Lemma 7.5A, then $\Omega = \Omega'$. Thus $f = g$ and the assertion of Theorem 6.5iii is trivially true.

If γ satisfies Lemma 7.5B, then by Proposition 7.4 there exist positive integers $n, d_0, d_1, \ldots, d_{n-1}$ such that

$$\gamma = \pm \begin{pmatrix} d_0 & 1 \\ 1 & 0 \end{pmatrix} \cdots \begin{pmatrix} d_{n-1} & 1 \\ 1 & 0 \end{pmatrix}.$$

By Lemma 7.2 and (8.6) we calculate that $|\Omega| = \gamma|\Omega'| = (d_0, \ldots, d_{n-1}, |\Omega'|)$. Noting that $(-1)^n = \det(\gamma) = \text{sign}(\Omega \cdot \Omega')$, we can apply Lemma 8.4 to conclude that $g = R^n f$. Thus the assertion of Theorem 6.5iii is true.

If γ satisfies Lemma 7.5C, then the argument of the previous paragraph applied to γ^{-1} proves that there is a positive integer n such that $f = R^n g$. Let $N > 0$ be such that $R^N g = g$ and let $k \in \mathbb{Z}$ be such that $kN > n$. Then

$g = R^{kN}g = R^{kN-n}(R^n g) = R^{kN-n}f$, and so in this case, too, the assertion of Theorem 6.5iii is verified.

The proof of Theorem 6.5iii is now complete. ∎

Exercises

1. Prove Lemma 8.3.

2. Let Δ be a nonsquare positive discriminant.

i. Let f be a reduced integral form of discriminant Δ. Let $\Omega_n = \Omega(f_n)$ and $\omega_n = \omega(f_n)$ for all $n \geq 0$. Let $k_n = [|\Omega_n|]$, so that $|\Omega_0| = (k_0, k_1, \ldots, k_{n-1}, |\Omega_n|)$ for all $n \geq 1$. Prove that $1/|\omega_n| = (k_{n-1}, k_{n-2}, \ldots, k_0, 1/|\omega_0|)$ for all $n \geq 1$.

ii. Keep the notation of part i. Prove that if $k_0 \geq 3$, then $|\Omega_0| + |\omega_0| > 3$. Prove that if $k_0 = 2$ and $k_n \leq 2$ for all $n \geq 0$, then $|\Omega_0| + |\omega_0| > 2 + \frac{1}{3}$. Prove that if $k_n = 1$ for all $n \geq 0$, then $|\Omega_0| + |\omega_0| = (\sqrt{5} + 1)/2 + (\sqrt{5} - 1)/2 = \sqrt{5}$.

iii. Prove that every proper equivalence class of integral forms of discriminant Δ contains a reduced form f such that $|\Omega f| + |\omega f| \geq \sqrt{5}$.

iv. Prove that every integral form of discriminant Δ is properly equivalent to a reduced form $[a, b, c]$ such that $|a| \leq \sqrt{\Delta/5}$. (This result sharpens Theorem 6.3i.)

3. i. Define a modified right neighbor $\mathscr{R}[a, b, c] = [a', b', c']$ for integral forms of positive nonsquare discriminant Δ by replacing Definition 6.1ii with the rule $b + b' \equiv 0 \pmod{2a'}$, $\sqrt{\Delta} - |2a'| < b' < \sqrt{\Delta} + |2a'|$, and $(b + \sqrt{\Delta})(b' - \sqrt{\Delta}) < 0$. Show that $\mathscr{R}f = Rf$ if f is reduced. Show that for all f, reduced or not, $\Omega(f)\Omega(\mathscr{R}f) < 0$, $|\Omega(\mathscr{R}f)| > 1$, and $|\Omega(f)| = (k_0, k_1, \ldots, k_{n-1}, |\Omega(\mathscr{R}^n f)|)$, where $k_i = [|\Omega(\mathscr{R}^i f)|]$ for all $i \geq 0$.

ii. Show that after replacing R by $\mathscr{R}$ and permitting d_0 to be zero, Lemma 8.4 becomes true for all forms f and g, reduced or not.

9. Automorphisms (II)

The reduction theory of Section 6 solves the problem of existence of proper equivalences between two forms f and g of the same positive nonsquare discriminant. In Section 9 we specialize to the case in which $f = g$, thereby deriving the theory of automorphisms of f by a method that is not dependent on a prior study of the Pell equation.

It is convenient to name some 2×2 matrices. We will write $T(\delta) = \begin{pmatrix} -\delta & 1 \\ -1 & 0 \end{pmatrix}$ and $S(k) = \begin{pmatrix} k & 1 \\ 1 & 0 \end{pmatrix}$ for all real numbers δ and k.

Definition 9.1. Let f be an integral form of positive nonsquare discriminant. We define $\tau_{f,n} \in \mathbf{SL}_2(\mathbb{Z})$ and $\sigma_{f,n} \in \mathbf{GL}_2(\mathbb{Z})$ for integers $n \geq 0$ by the following two formulas, where we have written $R^i f = [a_i, b_i, c_i]$ for all integers $i \geq 0$:

i. $\tau_{f,n} = T(\delta_0)T(\delta_1) \cdots T(\delta_{n-1})$, where $b_i + b_{i+1} = 2c_i\delta_i$.
ii. $\sigma_{f,n} = S(k_0)S(k_1) \cdots S(k_{n-1})$, where $k_i = |\delta_i|$.

The matrices $\sigma_{f,n}$ are really only of interest for reduced forms f.

Lemma 9.2. Let f be an integral form of positive nonsquare discriminant. Then $f = \tau_{f,n} R^n f$ for every integer $n \geq 0$.

Proof. The case $n = 1$ is the computation $Rf = \begin{pmatrix} 0 & -1 \\ 1 & -\delta_0 \end{pmatrix} f = T(\delta_0)^{-1} f$ that was mentioned immediately following Definition 6.1 of R. The general case follows by induction on n. ■

Lemma 9.3. Let f be a reduced integral form of positive nonsquare discriminant and let $\Omega = \Omega(f)$.

i. $|\Omega(f)| = (k_0, k_1, \ldots, k_{n-1}, |\Omega(R^n f)|)$, where the k_i are as in Definition 9.1ii, for all integers $n \geq 1$.

ii.
$$\tau_{f,n} = \begin{cases} (-1)^{n/2}\sigma_{f,n} & \text{if } \Omega > 0 \text{ and } n \text{ is even} \\ (-1)^{n/2}\begin{pmatrix} -1 & 0 \\ 0 & 1 \end{pmatrix}\sigma_{f,n}\begin{pmatrix} -1 & 0 \\ 0 & 1 \end{pmatrix} & \text{if } \Omega < 0 \text{ and } n \text{ is even} \\ (-1)^{(n-1)/2}\sigma_{f,n}\begin{pmatrix} -1 & 0 \\ 0 & 1 \end{pmatrix} & \text{if } \Omega > 0 \text{ and } n \text{ is odd} \\ (-1)^{(n+1)/2}\begin{pmatrix} -1 & 0 \\ 0 & 1 \end{pmatrix}\sigma_{f,n} & \text{if } \Omega < 0 \text{ and } n \text{ is odd.} \end{cases}$$

Proof. i. The proof of the implication $1 \Rightarrow 2$ of Lemma 8.4 actually proves this more precise statement.

ii. Since f is reduced, $\operatorname{sign}(\Omega) = \operatorname{sign}(\delta_0)$, and δ_i and δ_{i+1} have opposite signs for every i. Thus the result follows by induction from the calculation

$$T(\delta) = \begin{cases} S(|\delta|)\begin{pmatrix} -1 & 0 \\ 0 & 1 \end{pmatrix} & \text{if } \delta \geq 0 \\ \begin{pmatrix} 1 & 0 \\ 0 & -1 \end{pmatrix}S(|\delta|) & \text{if } \delta \leq 0 \end{cases}.$$
■

We proceed to the main theorem of Section 9.

Theorem 9.4. Let f be a reduced integral form of positive nonsquare discriminant. Let $N > 0$ be the smallest positive integer such that $R^N f = f$.

i. $\tau_{f,N}$ is an element of $\mathscr{A}ut^+(f)$ of infinite order.

ii. $\mathscr{A}ut^+(f) = \{\pm(\tau_{f,N})^m | m \in \mathbb{Z}\}$.

Proof. i. That $\tau_{f,N}$ is a proper automorphism of f is an immediate consequence of Lemma 9.2. Because $R^N f = f$, we have $(\tau_{f,N})^m = \tau_{f,mN}$ for every $m \geq 0 \in \mathbb{Z}$. An easy induction shows that $\tau_{f,n} \neq I$ for $n > 0$. Therefore, $\tau_{f,n}$ has infinite order.

ii. We have only to refine a bit the case $f = g$ of the proof of Theorem 6.5iii given in Section 8, to which we will refer.

Let f be reduced, let $\Omega = \Omega' = \Omega(f)$, and let $\tau \in \mathscr{A}ut^+(f)$. Define $\gamma \in \mathbf{SL}_2(\mathbb{Z})$ by (8.5). As in the proof of Theorem 6.5iii we can conclude that γ satisfies 7.5A, B, or C.

If γ satisfies 7.5A, then $\tau = \pm\begin{pmatrix} 1 & 0 \\ 0 & 1 \end{pmatrix}$.

Suppose that γ satisfies 7.5B, say $\gamma = \pm S(d_0)S(d_1) \cdots S(d_{n-1})$. As in the proof of Theorem 6.5iii we find that $|\Omega| = (d_0, \ldots, d_{n-1}, |\Omega|)$ and $R^n f = f$. Lemmas 7.3 and 9.3i imply that $d_i = k_i$ for all i; thus $\sigma_{f,n} = S(d_0)S(d_1) \cdots S(d_{n-1}) = \pm\gamma$. A comparison of (8.5) and Lemma 9.3ii shows that $\tau = \pm\tau_{f,n}$. The equation $R^n f = f$ together with the minimality property of N prove that $n = mN$ for some integer $m > 0$. Therefore, $\tau = \pm\tau_{f,mN} = \pm(\tau_{f,N})^m$.

If γ satisfies 7.5C, then the argument of the previous paragraph applied to γ^{-1} shows that $\tau^{-1} = \pm(\tau_{f,N})^m$ for some integer $m > 0$.

The proof of Theorem 9.4ii is complete. ■

Corollary 9.5. Let f be an integral binary quadratic form of positive nonsquare discriminant.

i. $\mathscr{A}ut^+(f) \simeq \{\pm 1\} \times \mathbb{Z}$.

ii. Let $M \geq 0$ be such that $R^m f$ is reduced and let $N > M$ be the smallest integer greater than M such that $R^N f = R^M f$.

Then $\mathscr{A}ut^+(f) = \{\pm(\tau_{f,N}\tau_{f,M}^{-1})^m | m \in \mathbb{Z}\}$.

Proof. i. If f is reduced, then the map ϕ: $\{\pm 1\} \times \mathbb{Z} \to \mathscr{A}ut^+(f)$ given by $\phi(\epsilon, m) = \epsilon(\tau_{f,N})^m$ with N as in Theorem 9.4 is an isomorphism.

If γf is reduced where $\gamma \in \mathbf{SL}_2(\mathbb{Z})$, then $\mathscr{A}ut^+(f) = \gamma^{-1}\mathscr{A}ut^+(\gamma f)\gamma \simeq \{\pm 1\} \times \mathbb{Z}$. This establishes Corollary 9.5i in general since for every f, by Theorem 6.3i, there exists $\gamma \in \mathbf{SL}_2(\mathbb{Z})$ such that γf is reduced.

ii. Take $\gamma = \tau_{f,M}^{-1}$, so that $\gamma f = R^M f$. The desired result follows from part i applied to $R^M f$ and the observation that $\tau_{f,M} \cdot \tau_{R^M f, N-M} = \tau_{f,N}$. ∎

As an example we determine $\mathscr{A}ut^+(f)$ for $f = 19X^2 - Y^2$. We must compute the sequence $R^n f = [a_n, b_n, c_n]$ until it begins to repeat, keeping track of the integers $\delta_n = (b_n + b_{n+1})/2c_n$, as follows.

n	0	1	2	3	4	5	6	7
a_n	19	−1	3	−5	2	−5	3	−1
b_n	0	8	4	6	6	4	8	8
c_n	−1	3	−5	2	−5	3	−1	3
δ_n	−4	2	−1	3	−1	2	−8	

The table shows that we may take $M = 1$, $N = 7$ in Corollary 9.5ii. Therefore, $\mathscr{A}ut^+(f) = \{\pm P^m | m \in \mathbb{Z}\}$, where

$$P = \tau_{f,7}(\tau_{f,1})^{-1} = \begin{pmatrix} -1421 & -170 \\ 326 & 39 \end{pmatrix} \begin{pmatrix} 4 & 1 \\ -1 & 0 \end{pmatrix}^{-1} = \begin{pmatrix} -170 & 741 \\ 39 & -170 \end{pmatrix}.$$

From Definition 5.2 and Theorem 5.3 we find that $\mathscr{P}ell(76) = \{\pm(-170, 39)^m | m \in \mathbb{Z}\}$, thus solving the Pell equation $X^2 - 19Y^2 = 1$ by reduction techniques.

Exercises

1. Let f be an integral binary form of positive nonsquare discriminant and let $n > 0$. Prove that $\tau_{f,n} \neq I$.

2. Show that

$$T(\delta_0)T(\delta_1) \cdots T(\delta_n) = \begin{pmatrix} p_n & p_{n-1} \\ q_n & q_{n-1} \end{pmatrix}, \quad \text{where} \begin{pmatrix} p_0 & p_{-1} \\ q_0 & q_{-1} \end{pmatrix} = \begin{pmatrix} -\delta_0 & 1 \\ -1 & 0 \end{pmatrix},$$

and $p_i = -(\delta_i p_{i-1} + p_{i-2})$ and $q_i = -(\delta_i q_{i-1} + q_{i-2})$ for $i \geq 1$.

3. Determine $\mathscr{A}ut^+(f)$ for $f = 9X^2 + 16XY + 6Y^2$; for $f = 31X^2 - Y^2$.

4. Let $f = [a, b, c]$ be a reduced integral form of positive nonsquare discriminant, let $N > 0$ be the smallest positive integer such that $|\Omega(R^n f)| = |\Omega(f)|$, and let $\mathscr{A}ut^*(f) = \{\gamma \in \mathbf{GL}_2(\mathbb{Z}) | \gamma f = \det(\gamma) f\}$, as in Exercise 5.4.

i. Show that $\tau_{f,N} f = f$ if N is even and that $\tau_{f,N}[a, -b, c] = -f$ if N is odd.

ii. Prove that $\mathscr{A}ut^*(f) = \mathscr{A}ut(f) = \{\pm(\tau_{f,N})^m | m \in \mathbb{Z}\}$ if N is even and that $\mathscr{A}ut^*(f) = \left\{\pm\left(\tau_{f,N}\begin{pmatrix} -1 & 0 \\ 0 & 1 \end{pmatrix}\right)^m | m \in \mathbb{Z}\right\}$ if N is odd.

5. Let f be an integral form of positive nonsquare discriminant. Let $M \geq 0$ be such that $R^M f$ is reduced and let $N > M$ be the smallest integer greater than M such that $|\Omega(R^N f)| = |\Omega(R^M f)|$.

i. Prove that $\mathscr{A}ut^*(f) = \left\{\pm\left(\tau_{f,N}\begin{pmatrix} \mu & 0 \\ 0 & 1 \end{pmatrix}\tau_{f,M}^{-1}\right)^m | m \in \mathbb{Z}\right\}$, where $\mu = (-1)^{N-M}$.

ii. Determine $\mathscr{A}ut^*(X^2 + XY - 18Y^2\}$. Compute the fundamental unit ϵ_{73}.

CHAPTER 5
The Class Group and Genera

1. Introduction

The final theorem of this book is another theorem of Gauss.

Theorem 1.1. Three Squares Theorem. Every positive integer m has a unique expression $m = 4^a u$, where a and u are integers and $4 \nmid u$. The following two conditions are equivalent.

1. $X^2 + Y^2 + Z^2 = m$ has a solution in integers X, Y, Z.
2. $u \not\equiv 7 \pmod 8$ (i.e., $u \equiv 1, 2, 3, 5$, or $6 \pmod 8$.)

Surprisingly, Gauss's proof of the Three Squares Theorem 1.1 depends largely on the theory of binary quadratic forms, the forms in just two variables we have long been studying. The key fact here was also discovered by Gauss.

Theorem 1.2. The set of proper equivalence classes of primitive integral binary quadratic forms of fixed nonzero discriminant Δ can be given the structure of finite abelian group in such a way that the following is true:

If two classes $\mathscr{C}_1$ and $\mathscr{C}_2$ represent integers m_1 and m_2, respectively, then the product class $\mathscr{C}_1\mathscr{C}_2$ represents the product $m_1 m_2$.

Theorem 1.2 suggests the existence of something like a homomorphism from the group of proper equivalence classes of primitive forms to the sets of integers represented by the forms in the classes. The working out of this idea amounts to Gauss's construction of the theory of genera of binary forms.

Roughly speaking, the sets of integers represented by the forms are replaced by the sets of congruence classes mod the discriminant to which the represented integers belong. Two primitive forms of the same discriminant Δ are in the same genus if they represent integers in the same congruence classes mod Δ. The genera can be described completely.

Let f and g be primitive forms of the same discriminant Δ that do not belong to the same genus. Then no prime number that is represented by f is congruent mod Δ to a prime that is represented by g. Should Δ be such that distinct equivalence classes of primitive forms of discriminant Δ belong to distinct genera, i.e., just one class in each genus, then the question of which primes can be represented by a given form of discriminant Δ can be answered completely with congruence conditions mod Δ. This is the case, for example, with $\Delta = -20$, where there are exactly two classes and two genera, represented by $X^2 + 5Y^2$ and $2X^2 + 2XY + 3Y^2$.

We will construct the class group of Theorem 1.2 in Section 2.

The facts of genus theory will be stated in Section 3. The proofs will be the concern of Sections 4–7. Along the way, in Section 4, we present Gauss's second proof of the Law of Quadratic Reciprocity.

The proof of Theorem 1.1 will come in Section 8, which is the final section of this book.

Exercises

1. Prove the implication $1 \Rightarrow 2$ of Theorem 1.1.

2. Let m be a positive integer. Deduce from Theorem 1.1 that if there exist $p, q, r \in \mathbb{Q}$ such that $p^2 + q^2 + r^2 = m$, then there exist $x, y, z \in \mathbb{Z}$ such that $x^2 + y^2 + z^2 = m$.

2. The Class Group

The theory to be presented rests on an identity that generalizes the product formula for sums of two squares. Though perhaps difficult to discover, it is easy to verify.

Basic Identity 2.1.

$$\left(a_1x_1^2 + bx_1y_1 + a_2cy_1^2\right)\left(a_2x_2^2 + bx_2y_2 + a_1cy_2^2\right)$$
$$= a_1a_2X^2 + bXY + cY^2,$$

where $X = x_1x_2 - cy_1y_2$ and $Y = a_1x_1y_2 + a_2y_1x_2 + by_1y_2$.

In the remainder of this section we will be concerned only with integral binary forms of a nonzero discriminant Δ, which we fix once for all. We will sometimes write $[a, b, c]$ for the form $aX^2 + bXY + cY^2$. The notation can be abbreviated to $[a, b, *]$ when $a \neq 0$, for then c is determined by the discriminant equation $\Delta = b^2 - 4ac$.

Definition 2.2. Two integral binary forms $f_1 = [a_1, b_1, c_1]$ and $f_2 = [a_2, b_2, c_2]$ of discriminant Δ are *concordant* iff the following three conditions are met.

i. $a_1 a_2 \neq 0$.
ii. $b_1 = b_2$.
iii. $a_2 | c_1$ and $a_1 | c_2$.

The *composition* $f_1 * f_2$ of two such concordant forms f_1 and f_2 of discriminant Δ is defined to be the form $[a_1 a_2, b, c]$, where $b = b_1 = b_2$ and $c = c_1/a_2 = c_2/a_1$. Since $f_1 * f_2$ has discriminant Δ, we can write $[a_1, b, *]*[a_2, b, *] = [a_1 a_2, b, *]$.

Note that condition i of the definition of concordant forms is automatically fulfilled in case the discriminant Δ is not a square. Moreover, condition iii follows from i and ii in case $\mathrm{GCD}(a_1, a_2) = 1$, since then the discriminant equation implies that $a_1 c_1 = a_2 c_2$.

The definitions of concordance and composition have been made just so that the following product rule will hold.

Proposition 2.3. If concordant forms f_1 and f_2 represent integers m_1 and m_2, respectively, then the composition form $f_1 * f_2$ represents the product $m_1 m_2$.

Proof. Immediate consequence of the Basic Identity 2.1. ■

Note that the composition of two primitive concordant forms is also a primitive form. It was one of Gauss's greatest discoveries that composition can be used to define a binary operation on the set of proper equivalence classes of primitive binary forms of fixed discriminant which makes the set into an abelian group. The remainder of Section 2 will show how this is done. We will write $f \sim g$ to indicate that two forms f and g are properly equivalent.

We begin with a useful lemma.

Lemma 2.4. Let $f = [a, b, c]$ be a primitive form and let M be a nonzero integer. Then f represents a nonzero integer that is relatively prime to M.

Proof. Write $2M = \pm\prod m_i \prod p_j \prod q_k$, where the m_i, p_j, q_k are primes such that $m_i \nmid a$, $p_j | a$, and $p_j \nmid c$, and $q_k | a$ and $q_k | c$. Let $r = \prod p_j$ and let $s = \prod m_i$. Using the fact that $q_k \nmid b$ because f is primitive, one sees that $\mathrm{GCD}(f(r, s), 2M) = 1$. ■

Lemma 2.5. Let $\mathscr{C}_1$ and $\mathscr{C}_2$ be proper equivalence classes of primitive forms of discriminant $\Delta \neq 0$. Let $M \neq 0 \in \mathbb{Z}$. Then there exists a pair of concordant forms $f_j = [a_j, b, *] \in \mathscr{C}_j$ such that $\mathrm{GCD}(a_1, a_2) = 1$ and $\mathrm{GCD}(a_1 a_2, M) = 1$.

Proof. To begin with, choose $F_1 = [a_1, b_1, *] \in \mathscr{C}_1$ such that $a_1 \neq 0$ and $\mathrm{GCD}(a_1, M) = 1$. To do this, let f be any element of $\mathscr{C}_1$. Let r, s be a pair of relatively prime integers such that $a_1 = f(r, s) \neq 0$ and $\mathrm{GCD}(a_1, M) = 1$. The pair r, s can be taken, for instance, as in the proof of Lemma 2.4. Let $t, u \in \mathbb{Z}$ be such that $\gamma = \left(\begin{smallmatrix} r & s \\ t & u \end{smallmatrix}\right) \in \mathbf{SL}_2(\mathbb{Z})$. Then $F_1 = \gamma f = [a_1, b_1, *]$ is as desired.

Similarly, choose $F_2 = [a_2, b_2, *] \in \mathscr{C}_2$ such that $a_2 \neq 0$ and $\mathrm{GCD}(a_2, a_1 M) = 1$.

Next find integers n_1, n_2 such that $b_1 + 2a_1 n_1 = b_2 + 2a_2 n_2$. This equation can be written $a_1 n_1 - a_2 n_2 = (b_2 - b_1)/2$. Solutions n_i exist because $b_1 \equiv \Delta \equiv b_2 \pmod 2$ and $\mathrm{GCD}(a_1, a_2) = 1$. It is clear that the forms

$$f_j = \begin{pmatrix} 1 & 0 \\ n_j & 1 \end{pmatrix} F_j = [a_j, b, *]$$

with $b = b_j + 2a_j n_j$ are concordant forms as demanded in the statement of the lemma. ∎

Proposition 2.6. Let $\mathscr{C}_1$ and $\mathscr{C}_2$ be proper equivalence classes of primitive forms of discriminant $\Delta \neq 0$. Let $f_j \in \mathscr{C}_j$ be a pair of concordant forms and let $g_j \in \mathscr{C}_j$ be a second pair of concordant forms. Then $f_1 * f_2$ is properly equivalent to $g_1 * g_2$.

Proof. Write $f_j = [a_j, b, c_j]$ and $g_j = [a'_j, b', c'_j]$.

The proof will advance in stages.

Stage 1. $f_1 = g_1$ and $\mathrm{GCD}(a_1, a'_2) = 1$.

Then f_1 is concordant with both f_2 and g_2, and we are to prove that $f_1 * f_2 \sim f_1 * g_2$.

Let $\gamma = \left(\begin{smallmatrix} r & s \\ t & u \end{smallmatrix}\right) \in \mathbf{SL}_2(\mathbb{Z})$ be such that $\gamma f_2 = g_2$. The equivalent matrix equation is

$$\gamma \cdot \begin{pmatrix} a_2 & b/2 \\ b/2 & c_2 \end{pmatrix} = \begin{pmatrix} a'_2 & b/2 \\ b/2 & c'_2 \end{pmatrix} \cdot {}^t\gamma^{-1}.$$

The above-diagonal component of this equation gives $-sc_2 = ta'_2$. Since f_1 is concordant with f_2, a_1 divides c_2. Thus $a_1|ta_2'$, whence it follows that $a_1|t$. Thus

$$\gamma' = \begin{pmatrix} r & sa_1 \\ t/a_1 & u \end{pmatrix} \in \mathbf{SL}_2(\mathbb{Z}).$$

Stage 1 is completed with the observation that $\gamma'(f_1 * f_2) = f_1 * g_2$.

Stage 2. $b = b'$ and $\mathrm{GCD}(a_1, a_2') = 1$.

The hypotheses of Stage 2 imply that f_1 and g_2 are concordant. Two applications of Stage 1 show that $f_1 * f_2 \sim f_1 * g_2 \sim g_1 * g_2$.

Stage 3. $\mathrm{GCD}(a_1 a_2, a_1' a_2') = 1$.

Let $B, n, n' \in \mathbb{Z}$ be such that $b + 2a_1 a_2 n = b' + 2a_1' a_2' n' = B$.

Set

$$F_1 = \begin{pmatrix} 1 & 0 \\ a_2 n & 1 \end{pmatrix} f_1 = [a_1, B, *]$$

and

$$F_2 = \begin{pmatrix} 1 & 0 \\ a_1 n & 1 \end{pmatrix} f_2 = [a_2, B, *] \in \mathscr{C}_2.$$

Let $H_1 = \left(\begin{smallmatrix} 1 & 0 \\ n & 1 \end{smallmatrix}\right)(f_1 * f_2) = [a_1 a_2, B, *]$. The discriminant equation applied to H_1 shows that $a_1 a_2 | (B^2 - \Delta)/4$. From the discriminant equations for F_1 and F_2 it then follows that F_1 and F_2 are concordant.

Similarly, the forms $G_j = [a_j', B, *] \in \mathscr{C}_j$ are concordant and $H_2 = [a_1' a_2', B, *] \sim g_1 * g_2$.

Stage 2 applies to the four forms $F_j, G_j \in \mathscr{C}_j$. We conclude that $f_1 * f_2 \sim H_1 = F_1 * F_2 \sim G_1 * G_2 = H_2 \sim g_1 * g_2$.

Final Stage. The General Statement.

By Lemma 2.5, there are concordant forms $F_j = [A_j, B, *] \in \mathscr{C}_j$ such that $\mathrm{GCD}(A_1 A_2, a_1 a_2 a_1' a_2') = 1$. Two applications of Stage 3 prove that $f_1 * f_2 \sim F_1 * F_2 \sim g_1 * g_2$. ■

By Lemma 2.5 and Proposition 2.6, composition of concordant forms gives a well-defined binary operation, called *composition* or *product*, on the set of proper equivalence classes of primitive forms of fixed nonzero discriminant. To be explicit, let $\mathscr{C}_1$ and $\mathscr{C}_2$ be two proper equivalence classes of primitive forms of discriminant $\Delta \neq 0$. Let $f_j \in \mathscr{C}_j$ be a pair of concordant forms. Then the composition $\mathscr{C}_1 \mathscr{C}_2$ of $\mathscr{C}_1$ and $\mathscr{C}_2$ is defined to be the proper equivalance class of the form $f_1 * f_2$.

***Definition* 2.7.** The *principal class* $\mathscr{C}_0$ of a nonzero discriminant Δ is the proper equivalence class of the *principal form* f_0 that is defined by the formula

$$f_0(X, Y) = \begin{cases} X^2 + XY + \dfrac{1-\Delta}{4}Y^2 & \text{if } \Delta \text{ is odd} \\ X^2 - \dfrac{\Delta}{4}Y^2 & \text{if } \Delta \text{ is even} \end{cases}.$$

***Definition* 2.8.** A *gaussian* form is an integral binary quadratic form of nonzero discriminant Δ that is primitive and, if $\Delta < 0$, is positive definite.

A proper equivalence class of forms is *gaussian* iff it contains gaussian forms.

Note that every form in a gaussian class is gaussian.

We can now state and prove the central result of this chapter.

Theorem 2.9. Let Δ be a nonzero discriminant.

i. The set of proper equivalence classes of primitive integral binary quadratic forms of discriminant Δ is a finite abelian group under composition of classes.

The identity element in the group is the principal class.

The inverse in the group of the class of a primitive form f is the class of any form that is improperly equivalent to f.

ii. The set of gaussian proper equivalence classes of discriminant Δ is a subgroup.

Proof. i. *Commutativity of composition of classes*. Clear from the Definition 2.2 of composition of concordant forms.

Identity element. Let $\mathscr{C}$ be a proper equivalence class as in the statement of the theorem and let $[a, b, *] \in \mathscr{C}$, where $a \neq 0$. Since $b \equiv \Delta \pmod 2$, we find that $f_0 \sim [1, b, *]$. Thus $\mathscr{C}_0\mathscr{C}$ is the class of $[1, b, *]*[a, b, *] = f \in \mathscr{C}$. Thus $\mathscr{C}_0$ is an identity element for composition of classes.

Inverses. Let $\mathscr{C}$ be a class. There is a form $[a, b, c] \in \mathscr{C}$ such that $ac \neq 0$, for we can take $a \neq 0$ by Lemma 2.4 and then consider $\begin{pmatrix} 1 & 0 \\ n & 1 \end{pmatrix}[a, b, *]$ for suitable n. Every form that is improperly equivalent to a form in $\mathscr{C}$ is properly equivalent to $\begin{pmatrix} 0 & 1 \\ 1 & 0 \end{pmatrix}[a, b, c] = [c, b, a]$, which is concordant with $[a, b, c]$. We compute the composition $[a, b, c]*[c, b, a] = [ac, b, 1] \sim f_0$. Thus the class of $[c, b, a]$ is inverse to the class $\mathscr{C}$.

Associativity. Let $\mathscr{C}_1$, $\mathscr{C}_2$, and $\mathscr{C}_3$ be any three classes. The trick here is to produce forms $f_j = [a_j, B, *] \in \mathscr{C}_j$ such that the coefficients a_j are nonzero and pairwise relatively prime. To do this, invoke Lemma 2.5 to begin with forms $g_j = [a_j, b_j, *] \in \mathscr{C}_j$ such that $a_1a_2a_3 \neq 0$ and $\mathrm{GCD}(a_1, a_2) =$

$\mathrm{GCD}(a_1a_2, a_3) = 1$. Then take

$$f_j = \begin{pmatrix} 1 & 0 \\ n_j & 1 \end{pmatrix} g_j,$$

where the integers n_j satisfy an equation $b_j + 2a_jn_j = B$ for some integer B independent of j. Such integers n_j always exist since the a_j are pairwise relatively prime and the b_j have the same parity (exercise). Finally we can compute $(f_1 * f_2) * f_3 = [a_1a_2, B, *]*[a_3, B, *] = [a_1a_2a_3, B, *] = f_1 * (f_2 * f_3)$. Thus composition of classes is associative.

Thus composition of classes is a group law. The group is finite by Corollary 8.4 and Exercise 8.8 of Chapter 2 and Corollary 6.4 of Chapter 4.

ii. There is nothing to prove unless $\Delta < 0$. The result for negative discriminants is clear from Definition 2.2 since then $[a, b, *]$ is positive definite if and only if $a > 0$. ■

Definition 2.10. The *class group* $\mathcal{C}\ell(\Delta)$ is defined for nonzero discriminants Δ to be the group of gaussian proper equivalence classes of discriminant Δ under composition of classes.

The proofs of Lemmas 2.4 and 2.5 actually give a method for calculating the composition of classes. We close this section with an example. We determine $\mathcal{C}\ell(-39)$.

By Theorem 8.7 of Chapter 2 the elements of $\mathcal{C}\ell(-39)$ will correspond to the reduced gaussian forms of discriminant -39. It is easy to list them all: $f_0 = [1, 1, 10]$, $f_1 = [2, 1, 5]$, $f_2 = [3, 3, 4]$, and $f_3 = [2, -1, 5]$. Thus $\mathcal{C}\ell(-39)$ is a group of order 4.

Let $\mathcal{C}_j$ be the proper equivalence class of f_j. Since $f_3 = \begin{pmatrix} 1 & 0 \\ 0 & -1 \end{pmatrix} f_1$, f_1 and f_3 are improperly equivalent. Hence $\mathcal{C}_1^{-1} = \mathcal{C}_3$. Since $\mathcal{C}_1$ is not self-inverse it cannot have order 2. Thus $\mathcal{C}_1$ is a generator $\mathcal{C}\ell(-39)$, which is cyclic: $\mathcal{C}_1^j = \mathcal{C}_j$.

Alternatively, compute $\mathcal{C}_1^2$ directly: $[2, 1, 5] \sim [5, -1, 2]$. $\begin{pmatrix} 1 & 0 \\ 2 & 1 \end{pmatrix}[2, 1, *] = [2, 9, *]$ and $\begin{pmatrix} 1 & 0 \\ 1 & 1 \end{pmatrix}[5, -1, *] = [5, 9, *]$. Thus $\mathcal{C}_1^2$ is the class of $[2, 9, *]*[5, 9, *] = [10, 9, *] \sim f_2$. Therefore, $\mathcal{C}_1^2 = \mathcal{C}_2$.

Exercises

1. Let $\Delta = b^2 - 4a_1a_2c$. Express $(a_1x_1 + ((b + \sqrt{\Delta})/2)y_1)(a_2x_2 + ((b + \sqrt{\Delta})/2)y_2)$ in the form $a_1a_2X + ((b + \sqrt{\Delta})/2)Y$. Compare with the Basic Identity 2.1.

2. Prove that there is just one proper equivalence class of forms of a given discriminant that represents 1, namely the principal class.

3. i. Let M and N be positive integers. Write $N = \prod p_i \prod q_j$, where the p_i and q_j are primes such that $p_i | M$ and $q_j \nmid M$. Let $N_0 = N$ and let $N_{k+1} = N_k / \mathrm{GCD}(N_k, M)$ for $k \geq 0$. Show that there exists k such that $N_k = N_{k+1}$, from which it follows that $N_k = \prod q_j$.

ii. Show that the integers r and s in the proof of Lemma 2.4 can be found from the data M, a, c without factoring.

4. Prove the existence of the integers n_j appearing in the proof of associativity in Theorem 2.9i.

5. Prove that $X^2 + 5Y^2$ represents the product mn of any two integers m and n that are represented by $2X^2 + 2XY + 3Y^2$.

6. Prove that $X^2 + XY + 4Y^2$ represents the product mn of any two integers m and n that are represented by $2X^2 + XY + 2Y^2$.

7. List all reduced gaussian forms of discriminant -55. Determine the group structure on this set.

8. Determine $\mathscr{C}\ell(\Delta)$ for $\Delta = -84, -47, -87, -95, -163$.

9. i. Let Δ be a nonzero discriminant and let $\mathscr{C}_1, \mathscr{C}_2 \in \mathscr{C}\ell(\Delta)$. Let $f_j = [a_j, b_j, *] \in \mathscr{C}_j$ and let $F \in \mathscr{C}_1\mathscr{C}_2$. Show that there exist 2×2 matrices $A = (a_{ij})$ and $B = (b_{ij})$ with entries $a_{ij}, b_{ij} \in \mathbb{Z}$ such that $f_1(x, y)f_2(u, v) = F(X, Y)$, where $X = (x \;\; y)A\binom{u}{v} = a_{11}xu + a_{12}xv + a_{21}yu + a_{22}yv$, $Y = (x \;\; y)B\binom{u}{v}$, $\det\begin{pmatrix} a_{11} & a_{12} \\ b_{11} & b_{12} \end{pmatrix} = a_1$, and $\det\begin{pmatrix} a_{11} & b_{11} \\ a_{21} & b_{21} \end{pmatrix} = a_2$. (Note: Gauss actually *defines* a form F to be a composition of the forms f_1 and f_2 iff there exist A and B giving such an identity relating f_1, f_2, and F. This definition is natural and quite important, but it is somewhat difficult to show that it leads to a group law on $\mathscr{C}\ell(\Delta)$.)

ii. Find A and B as in i for

a. $f_1 = f_2 = 2X^2 + XY + 2Y^2$ and $F = X^2 + XY + 4Y^2$.
b. $f_1 = f_2 = 2X^2 + 2XY + 3Y^2$ and $F = X^2 + 5Y^2$.
c. $f_1 = f_2 = 2X^2 + XY + 3Y^2$ and $F = 2X^2 - XY + 3Y^2$.

10. Let k be an integer greater than 3 that is congruent to 3 (mod 8). Prove that $[4, 2, (k+1)/4] \in \mathscr{C}\ell(-4k)$ has order 3.

11. Prove that $[2, 1, k] \in \mathscr{C}\ell(1 - 8k)$ has order at least 5 for $k \geq 9$. (*Suggestion*: Show that $[2, 1, 2r]^2 \sim [4, 1, r]$ and that $[2, 1, 2r + 1]^2 \sim [4, -3, r + 1]$.) What is the order for $1 \leq k \leq 8$?

12. Let Δ be a negative discriminant. Show that the group of all proper equivalence classes of primitive forms of discriminant Δ is isomorphic to $\{\pm 1\} \times \mathscr{C}\ell(\Delta)$.

13. Let m be a positive integer. Prove that the map $f(aX^2 + mXY) = \bar{a}$ induces an isomorphism $f\colon \mathscr{C}\ell(m^2) \to U_m$. (See Exercise 8.8 of Chapter 2.)

14. Show that composition is a well-defined binary operation on equivalence classes of primitive forms of discriminant $\Delta \neq 0$ (as opposed to proper equivalence classes) if and only if $\mathscr{C}^2 = \mathscr{C}_0$ for every $\mathscr{C} \in \mathscr{C}\ell(\Delta)$.

15. **i.** Let $n \neq 0 \in \mathbb{Z}$ and let $(u_1, u_2, u_3), (v_1, v_2, v_3) \in \mathbb{Z}^3$. Suppose that $\mathrm{GCD}(u_1, u_2, u_3) = 1$ and $u_i v_j \equiv u_j v_i \pmod{n}$ for all i and j. Prove that there exists $x \in \mathbb{Z}$ such that $u_i x \equiv v_i \pmod{n}$ for all $i = 1, 2, 3$.

ii. Let $\Delta \neq 0$ be a discriminant and let $f_i = [a_i, b_i, c_i]$, $i = 1, 2$, be two integral forms of discriminant Δ. Suppose that $a_1 a_2 \neq 0$ and that $\mathrm{GCD}(a_1, a_2, (b_1 + b_2)/2) = 1$. Prove that there exist integers n_1 and n_2 such that

$$\begin{pmatrix} 1 & 0 \\ n_1 & 1 \end{pmatrix} f_1 \quad \text{and} \quad \begin{pmatrix} 1 & 0 \\ n_2 & 1 \end{pmatrix} f_2$$

are concordant. (*Hint*:

$$\begin{pmatrix} 1 & 0 \\ n_i & 1 \end{pmatrix} f_i = [a_i, B, *],$$

where $a_2 B \equiv a_2 b_1$, $a_1 B \equiv a_1 b_2$, and $((b_1 + b_2)/2)B \equiv (\Delta + b_1 b_2)/2 \pmod{2a_1 a_2}$.)

16. Let Δ be a negative discriminant and let $p < |\Delta|/4$ be a prime number.

i. Show that p is not represented by the principal form f_0 of discriminant Δ.

ii. Prove that $|\mathscr{C}\ell(\Delta)| > 1$ if either (a) $p \nmid \Delta$ and $\chi_\Delta(p) = 1$ or (b) $p | \Delta$ and $p^2 \nmid \Delta$ or (c) $|\Delta| > 8$ and $\Delta \equiv 8$ or $12 \pmod{16}$.

17. Let Δ be a negative discriminant such that $|\mathscr{C}\ell(\Delta)| = 1$. Show that:

$$\begin{array}{ll}
\text{if } \Delta < -8 & \text{then } \Delta \equiv 0, 4, 5, \text{ or } 13 \pmod{16} \\
\text{if } \Delta < -12 & \text{then } \Delta \equiv 2 \pmod{3} \text{ or } 0 \pmod{9} \\
\text{if } \Delta < -20 & \text{then } \Delta \equiv 2 \text{ or } 3 \pmod{5} \text{ or } 0 \pmod{25} \\
\text{if } \Delta < -28 & \text{then } \Delta \equiv 3, 5, \text{ or } 6 \pmod{7} \text{ or } 0 \pmod{49}.
\end{array}$$

18. For $x > 0$ let $\tau(x)$ equal the number of discriminants Δ such that $-x < \Delta < 0$ and $|\mathscr{C}\ell(\Delta)| = 1$. Prove that $\lim_{x\to\infty}\tau(x)/x = 0$. (See Exercise 16.) In fact it is known that there are exactly 13 negative discriminants Δ such that $|\mathscr{C}\ell(\Delta)| = 1$.

3. The Genus Group

Definition 3.1. An equivalence class or proper equivalence class of integral forms is said to *represent* an integer m iff m is represented by the forms in the class.

The central problem of this chapter is to describe the set of primes that are represented by an equivalence class of forms. With the organization of the proper equivalence classes of gaussian forms of fixed nonzero discriminant Δ into the class group $\mathscr{C}\ell(\Delta)$ we have already constructed our main tool. It remains but to use it.

The product rule Proposition 2.3 asserts that the composition $\mathscr{C}_1\mathscr{C}_2$ of two primitive classes $\mathscr{C}_1$ and $\mathscr{C}_2$ will represent the product m_1m_2 of any two integers m_j represented by the $\mathscr{C}_j$. Thus the map that associates to each element of $\mathscr{C}\ell(\Delta)$ the set of integers that that class represents has a homomorphism-like property. A slight modification produces an actual group homomorphism. We will associate to each element of $\mathscr{C}\ell(\Delta)$ certain congruence classes mod Δ that contain integers represented by the class.

Definition 3.2. For a nonzero discriminant Δ define the subgroup H_Δ of U_Δ to be the set of all $\bar{x} \in U_\Delta$ such that

i. $(x/p) = +1$ for all odd prime divisors p of Δ and

ii. $$x \equiv \begin{cases} 1 \bmod 4 & \text{if } \Delta \equiv 12 \bmod 16 \text{ or } 16 \bmod 32 \\ 1 \bmod 8 & \text{if } \Delta \equiv 0 \bmod 32 \\ 1 \text{ or } 7 \bmod 8 & \text{if } \Delta \equiv 8 \bmod 32 \\ 1 \text{ or } 3 \bmod 8 & \text{if } \Delta \equiv 24 \bmod 32 \end{cases}.$$

Theorem 3.3. Let Δ be a nonzero discriminant.

i. Let $\mathscr{C} \in \mathscr{C}\ell(\Delta)$. Let m, n be integers that are represented by $\mathscr{C}$ and are relatively prime to Δ. Then $\overline{m} = \overline{n} \in U_\Delta/H_\Delta$.

ii. Let ω_Δ: $\mathscr{C}\ell(\Delta) \to U_\Delta/H_\Delta$ be the function defined by $\omega_\Delta(\mathscr{C}) = \overline{m}$, where $\mathscr{C}$ and m are as in *i*. Then ω_Δ is a group homomorphism.

Proof. i. Write $m \equiv nx \pmod{\Delta}$. We are to prove that $\bar{x} \in H_\Delta$. Since $mn \equiv xn^2 \pmod{\Delta}$ and $\bar{n}^2 \in H_\Delta$ (because H_Δ contains all squares), it will suffice to prove that $\overline{mn} \in H_\Delta$.

Let f be a form in $\mathscr{C}$, and let $r, s, t, u \in \mathbb{Z}$ be such that $f(r, s) = m$ and $f(t, u) = n$. Let $\gamma = \begin{pmatrix} r & s \\ t & u \end{pmatrix}$. Then $\gamma f = [m, l, n]$ for some integer l. Equating discriminants gives $l^2 - 4mn = \Delta X^2$, where $X = \det \gamma$. For every odd prime divisor p of Δ we have the congruence $4mn \equiv l^2 \pmod{p}$, which shows that $(mn/p) = +1$. Thus mn satisfies 3.2i.

Now suppose that Δ is even. We want to check that mn satisfies condition 3.2ii. Note that l is even and let $L = l/2$. Then $mn = L^2 - \Delta X^2/4$. Let $\Delta = 2^a d$, where d is odd. Since mn is odd, L is odd in case $a \geq 3$. All now follows from separate consideration of the four congruences:

$$mn \equiv \begin{cases} L^2 - dx^2 \pmod{4} & \text{if } a = 2 \\ 1 - 2dX^2 \pmod{8} & \text{if } a = 3 \\ 1 - 4dX^2 \pmod{8} & \text{if } a = 4 \\ 1 \qquad\qquad \pmod{8} & \text{if } a \geq 5 \end{cases}.$$

ii. The existence of an integer m as in Theorem 3.3i is assured by Lemma 2.4. Theorem 3.3i proves that the function ω_Δ is well defined. That ω_Δ is a group homomorphism follows trivially from the product rule Proposition 2.3. ∎

Definition 3.4. The *gauss symbol* ω_Δ is defined for nonzero discriminants Δ to be the homomorphism ω_Δ: $\mathscr{C}\ell(\Delta) \to U_\Delta/H_\Delta$ that is constructed in Theorem 3.3.

Definition 3.5. The *genus group* (*group of genera*) $\mathscr{G}en(\Delta)$ is defined for nonzero discriminants Δ to be the quotient group $\mathscr{G}en(\Delta) = \mathscr{C}\ell(\Delta)/(\ker \omega_\Delta)$.

The identity element $\mathscr{G}_0$ in $\mathscr{G}en(\Delta)$ is called the *principal genus*. Thus a gaussian proper equivalence class $\mathscr{C}$ and the forms it contains are said to belong to the principal genus iff $\omega_\Delta(\mathscr{C}) = 1$.

Explicit computation of the Gauss symbol ω_Δ for a nonsquare Δ yields information on the representability of prime numbers by forms of discriminant Δ which sometimes refines that given by the Kronecker symbol χ_Δ. Gauss determined both the image and kernel of ω_Δ in brilliant sections of his *Disquisitiones*. We conclude this section with an introduction to these topics and an example.

Let Δ be a nonsquare discriminant. The kernel of the Kronecker symbol χ_Δ contains the congruence classes of primes that can be represented by primitive forms of discriminant Δ. The image of the gauss symbol ω_Δ is given by the congruence classes of integers that can be represented by primitive forms of discriminant Δ. It is perhaps not surprising that the image of ω_Δ and the kernel of χ_Δ are equal. That is our next result.

Theorem 3.6. Let Δ be a nonsquare discriminant.

i. $H_\Delta \subset \ker \chi_\Delta$.

ii. $\operatorname{im} \omega_\Delta = (\ker \chi_\Delta)/H_\Delta$.

iii. $\mathcal{G}en(\Delta) \simeq (\ker \chi_\Delta)/H_\Delta$.

Proof. i. Let $\Delta = 2^a d$, where d is odd and let $\bar{x} \in H_\Delta$. The condition 3.2i of the definition of H_Δ evaluates the Jacobi symbol $(x/|d|) = 1$.

If Δ is odd, then $\chi_\Delta(x) = (x/|d|) = 1$.

If Δ is even, then

$$\chi_\Delta(x) = (-1)^{(d-1)/2\cdot(x-1)/2}(2/|x|)^a(x/|d|) = (-1)^{(d-1)/2\cdot(x-1)/2}(2/|x|)^a.$$

It is easy to check, using Definition 3.2ii, that both factors in the final product for $\chi_\Delta(x)$ equal 1 except when $\Delta \equiv 24 \pmod{32}$ and $x \equiv 3 \pmod 8$, in which case both factors equal -1. In all cases, $\chi_\Delta(x) = 1$.

ii. We prove first that $\operatorname{im} \omega_\Delta \subset (\ker \chi_\Delta)/H_\Delta$.

Let $\mathscr{C} \in \mathscr{C}\ell(\Delta)$ and let m be an odd integer relatively prime to Δ that is represented by $\mathscr{C}$. We can assume that $m > 0$. If $\Delta < 0$, this is automatic because then $\mathscr{C}$ is positive definite. If $\Delta > 0$, then it is easy to see that the principal class in $\mathscr{C}\ell(\Delta)$ represents negative odd integers relatively prime to Δ and so the product rule Proposition 2.3 can be used to insure that $m > 0$. Let $f \in \mathscr{C}$ and let $r, s \in \mathbb{Z}$ be such that $f(r, s) = m$. After replacing r, s, and m by r/g, s/g, and m/g^2, where $g = \operatorname{GCD}(r, s)$, we may assume that $\operatorname{GCD}(r, s) = 1$.

Let $\gamma = \begin{pmatrix} r & s \\ t & u \end{pmatrix} \in \mathbf{SL}_2(\mathbb{Z})$. Let $\gamma f = [m, l, n]$. The discriminant equation gives $\Delta = l^2 - 4mn \equiv l^2 \pmod m$. Thus $\chi_\Delta(m) = (\Delta/m) = (l/m)^2 = 1$. Therefore, $\omega_\Delta(\mathscr{C}) = \bar{m} \in \ker \chi_\Delta$.

Finally we prove that $\ker \chi_\Delta \subset \operatorname{im} \omega_\Delta$. Let $\bar{m} \in \ker \chi_\Delta$. By Dirichlet's Theorem on Primes in Arithmetic Progressions, there is a prime number p such that $p \equiv m \pmod \Delta$. Since $\chi_\Delta(p) = 1$, by Theorem 10.1 of Chapter 3 there is a form f of discriminant Δ that represents p. Since $p \nmid \Delta$, by Proposition 10.4 of Chapter 3 the form f is primitive. Thus $\bar{m} = \bar{p} = \omega_\Delta(\mathscr{C}) \in \operatorname{im} \omega_\Delta$, where $\mathscr{C}$ is the proper equivalence class of f.

iii. Fundamental Theorem of Homomorphisms. ∎

The inclusion $(\ker \chi_\Delta)/H_\Delta \subset \operatorname{im} \omega_\Delta$ of Theorem 3.6ii is a very strong existence theorem for genera. It will play an essential part in the proof of the Three Squares Theorem 1.1. A proof of this inclusion that does not depend on Dirichlet's Theorem on Primes in Arithmetic Progressions will be discussed in Section 4.

Definition 3.7. The homomorphism sq_Δ: $\mathscr{C}\ell(\Delta) \to \mathscr{C}\ell(\Delta)$ is defined for nonzero discriminants Δ by the formula $sq_\Delta(\mathscr{C}) = \mathscr{C}^2$.

The group of *ambiguous classes* $\mathscr{A}mb(\Delta)$ is defined to be the kernel of sq_Δ.

The group $\mathscr{S}q(\Delta)$ is defined to be the image of sq_Δ. Thus $\mathscr{S}q(\Delta) = \{\mathscr{C}^2 | \mathscr{C} \in \mathscr{C}\ell(\Delta)\}$.

Theorem 3.8. The Duplication Theorem. Let Δ be a nonzero discriminant. Then $\ker \omega_\Delta = \mathscr{S}q(\Delta)$.

Start of Proof. The inclusion $\mathscr{S}q(\Delta) \subset \ker \omega_\Delta$ is trivial because every element of U_Δ/H_Δ has order 1 or 2. The proof of the reverse inclusion requires quite different techniques. We begin the discussion of these in the next section. The proof of Theorem 3.8 will be completed in Section 7. ■

We illustrate the preceding theory by continuing the example of discriminant -39 which we began in Section 2.

By Theorem 2.6 of Chapter 3, we discover that $H_{-39} = \{x^2 | x \in U_{39}\}$. From this it is easy to show that $H_{-39} = \{1, 4, 10, 16, 22, 25\} \subset U_{39}$. Since a form $[a, b, c]$ represents both a and c, one observes that $\omega_{-39}(\mathscr{C}_0) = \omega_{-39}(\mathscr{C}_2) = H_{-39}$ and that $\omega_{-39}(\mathscr{C}_1) = \omega_{-39}(\mathscr{C}_3) = 2H_{-39} = \{2, 5, 8, 11, 20, 32\} \subset U_{39}$. Note that these calculations verify Theorems 3.6 and 3.8 in the case $\Delta = -39$.

Now let p be a prime number other than 3 or 13. By Theorem 10.1 of Chapter 3, p is represented by a form of discriminant -39 if and only if $\chi_{-39}(p) = 1$ if and only if p is congruent mod 39 to an element of $H_{-39} \cup 2H_{-39}$. Taking into account the fact that f_1 and f_3 are equivalent (though not properly equivalent) and hence represent the same integers, we learn that p is represented by f_0 or f_2 if and only if $p \equiv 1$, 4, 10, 16, 22, or 25 (mod 39) and that p is represented by f_1 if and only if $p \equiv 2, 5, 8, 11, 20, 32 \pmod{39}$. Finally note that 3 and 13 are represented by f_2.

We have not given a congruence condition that determines which of the two forms f_0 or f_2 represents a prime $p \in H_{-39}$, and in fact it can be proven that no such congruence condition exists. The problem is that $\omega_{-39}(\mathscr{C}_0) = \omega_{-39}(\mathscr{C}_2)$. The situation is nicer for discriminants Δ for which ω_Δ is injective, i.e., for discriminants such that there is just one equivalence class of forms in each genus. But even for $\Delta = -39$ we have gone beyond the information available from the Kronecker symbol.

Exercises

1. Which prime numbers can be represented by $X^2 + 5Y^2$? Answer with congruence conditions. (Compare with Exercise 10.7 of Chapter 3.)

2. List the reduced gaussian forms of discriminant $\Delta = -15, -55, -84, -23$. Determine as far as possible with congruences which primes each of these forms represents.

3. Find a nonzero discriminant Δ and a positive integer $m \in H_\Delta$ such that m is not represented by any form of discriminant Δ.

4. **i.** Let m be a positive integer. Show that there is a homomorphism $\omega: U_m \to U_{m^2}/H_{m^2}$ such that $\omega(\bar{a}) = \bar{a}$. Prove that ω is surjective and that $\ker \omega = \{x^2 | x \in U_m\}$.

ii. Let Δ be a nonzero square. Prove that $\omega_\Delta: \mathscr{C}\ell(\Delta) \to U_\Delta/H_\Delta$ is surjective and that $\ker \omega_\Delta = \mathscr{S}q(\Delta)$. (See Exercise 2.13.)

5. Show that the class group $\mathscr{C}\ell(\Delta)$ of a nonsquare discriminant $\Delta = \pm p^a q^b$, p and q odd primes, cannot be isomorphic to $\{\pm 1\} \times \{\pm 1\}$.

6. **i.** Let $\Delta \equiv 1 \pmod 4$. Prove that $H_\Delta = \{x^2 | x \in U_\Delta\}$.

ii. Prove that $H_{4n} = \{x^2, x^2(1-n) | x \in U_{4n}\} = \{x^2, x^2 - n | x \in U_{4n}\}$ if n is even and that $H_{4n} = \{x^2, x^2(4-n) | x \in U_{4n}\} = \{x^2, 4x^2 - n | x \in U_{4n}\}$ if n is odd.

7. Let $a, b \in \mathbb{Z}$ be nonzero and relatively prime. Prove that the following three assertions are equivalent. (a) There exist $\alpha, \beta \in \mathbb{Z}$ such that $\alpha^2 \equiv a \pmod b$ and $\beta^2 \equiv b \pmod a$, and a and b are not both negative. (b) $aX^2 + bY^2$ is in the principal genus of $\mathscr{C}\ell(-4ab)$. (c) There exist integers x, y, z such that $ax^2 + by^2 = z^2$ and $\mathrm{GCD}(z, 4ab) = 1$.

8. Let Δ be a nonzero discriminant. Let $\mathscr{C}\ell(\Delta) \simeq A \oplus C(2)^r \oplus_{i=1}^{s} C(2^{m_i})$, where A has odd order, $C(n)$ denotes a cyclic group of order n, $m_i \geq 2$, and $r, s \geq 0$. Prove that $|\mathscr{A}mb(\Delta)| = 2^{r+s}$ and that $|\mathscr{A}mb(\Delta) \cap \mathscr{G}_0| = 2^s$.

9. Let Δ be a negative discriminant and let $\mathscr{C}\ell^*(\Delta)$ be the group of all proper equivalence classes of primitive forms of discriminant Δ, positive definite or not. Show that a homomorphism $\omega_\Delta^*: \mathscr{C}\ell^*(\Delta) \to U_\Delta/H_\Delta$ which extends ω_Δ may be defined by the same procedure that defines ω_Δ. Prove that ω_Δ^* is surjective.

10. A positive integer d is said to be *convenient* iff ω_{-4d} is injective. (There are 65 known convenient numbers, all found by Euler. There are probably no more.)

i. Prove that 3, 7, and 15 are the only convenient numbers that are congruent to 3 (mod 4). (*Hint*: If $d + 1 = 2^r c$ where c is odd and greater than 1, consider $[2^r, 2, c]$. If $d + 1 = 2^r$, consider $[8, 6, 2^{r-3} + 1]$.)

ii. Let d be a convenient number that is congruent to 1 (mod 4). Prove that $d = 1$ or $d + 1 = 2p$, where p is an odd prime number. Prove that $d + 4 = q$ or q^2, where q is an odd prime number.

iii. Find all odd convenient numbers less than 100. (There are 15 of them.)

11. Let d be an even convenient number.

i. Prove that $d + 1 = p$ or p^2, where p is an odd prime number.

ii. Prove that $d = 18$ or 72 or $d + 9 = q$, q^2, or $3q$, where q is a prime number greater than 3. (*Hint*: Consider separately the cases $3 \mid d$ and $3 \nmid d$. Eliminate the possibility $d + 9 = 3^r$, $r \geq 6$, with $[27, 12, 3^{r-3} + 1]$.)

iii. Find all even convenient numbers less than 100. (There are 21 of them.)

12. Let d be a positive odd integer such that $-d$ is a discriminant and ω_{-d} is injective.

i. Show that if $d \equiv 7 \pmod 8$, then $d = 7$ or 15.

ii. Show that if $d \equiv 3 \pmod 8$, then $d = 3$ or $d + 1 = 4p$ or $4p^2$, where p is an odd prime number.

iii. Find all odd negative discriminants Δ such that $|\Delta| < 100$ and ω_Δ is injective. (There are 13 of them.)

13. List the reduced gaussian forms of discriminants -144 and -256. Determine as far as possible with congruences which primes each of these forms represents. Then turn to Exercises 10.8 and 8.6 of Chapter 3.

4. What Gauss Did

Throughout this section Δ will denote a nonsquare discriminant.

We want to discuss the relationship between the following three true propositions, each of which has a direct proof.

Proposition 4.1. $\operatorname{im}\omega_\Delta = (\ker \chi_\Delta)/H_\Delta$.

We have already stated and proved Proposition 4.1 as Theorem 3.6. The proof of the inclusion $(\ker \chi_\Delta)/H_\Delta \subset \operatorname{im}\omega_\Delta$ relied crucially on Dirichlet's Theorem on Primes in Arithmetic Progressions, which we have not proved.

Proposition 4.2. $\ker\omega_\Delta = \operatorname{im} sq_\Delta$.

This is Gauss's Duplication Theorem 3.8. The most natural proofs of the containment $\ker\omega_\Delta \subset \operatorname{im} sq_\Delta$ rest on the theory of integral quadratic forms in three variables. We will present such a proof of Proposition 4.2 in Section 7.

Proposition 4.3. $|\mathscr{A}m\ell(\Delta)| = \frac{1}{2}|U_\Delta/H_\Delta|$.

Proposition 4.3 can be proved by separate evaluation of the two sides of the equation. This will be the subject of Section 5.

Gauss discovered that any two of the preceding three propositions easily imply the third. Dirichlet's Theorem was not available to Gauss in the 1790s, so he proved Proposition 4.1 as a consequence of Propositions 4.2 and 4.3.

Demonstration That Any Two of the Propositions 4.1–4.3 Easily Imply the Third. To begin with, we regard the two inclusions $\operatorname{im}\omega_\Delta \subset (\ker \chi_\Delta)/H_\Delta$ and $\operatorname{im} sq_\Delta \subset \ker\omega_\Delta$ as easy. Elementary proofs were given in Section 3. Therefore Propositions 4.1 and 4.2 will follow from the weaker assertions:

Proposition 4.1′. $|\operatorname{im}\omega_\Delta| = |(\ker \chi_\Delta)/H_\Delta|$.

Proposition 4.2′. $|\ker\omega_\Delta| = |\operatorname{im} sq_\Delta|$.

We must now show that any two of Propositions 4.1′, 4.2′, and 4.3 imply the third. This is a simple exercise starting with three known equalities (4.4). Both (4.4a) and (4.4b) are versions of the Fundamental Theorem of Homomorphisms. Note that (4.4c) is a consequence of Theorem 3.6i and Proposition 9.3vi of Chapter 3.

$$|\ker\omega_\Delta| \cdot |\operatorname{im}\omega_\Delta| = |\mathscr{C}\ell(\Delta)|, \tag{4.4a}$$

$$|\mathscr{A}m\ell(\Delta)| \cdot |\operatorname{im} sq_\Delta| = |\mathscr{C}\ell(\Delta)|, \tag{4.4b}$$

$$|(\ker \chi_\Delta)/H_\Delta| = \tfrac{1}{2}|U_\Delta/H_\Delta|. \quad \blacksquare \tag{4.4c}$$

Since $C\ell(\Delta)$ is a finite abelian group, there is an isomorphism $\mathscr{C}\ell(\Delta) \simeq A \oplus_{i=1}^{r} C(2^{n_i})$, where A had odd order, $C(n)$ denotes a cyclic group of order

n, all $n_i > 0$, and $r \geq 0$. Then $|\mathcal{A}mb(\Delta)| = 2^r$. Therefore, an evaluation of $|\mathcal{A}mb(\Delta)|$ as in Proposition 4.3 gives some information on the structure of the class group $\mathcal{C}\ell(\Delta)$, about which very little is known in general.

Gauss observed that the Law of Quadratic Reciprocity Proposition 1.2 of Chapter 3 can be deduced from Proposition 4.3 alone. We close this section with Gauss's second proof of quadratic reciprocity. It is very satisfying to have a proof of the reciprocity law that springs from the theory of binary quadratic forms.

Proof of Proposition 1.2 of Chapter 3. We prove the equivalent statement Proposition 3.4iii. We will need only a weakened form of Proposition 4.3, namely:

Proposition 4.3′. $|\mathcal{A}mb(\Delta)| \leq \frac{1}{2}|U_\Delta/H_\Delta|$.

The three assertions Proposition 4.3′, (4.4a), and (4.4b) are easily combined with the trivial inclusion $\text{im } sq_\Delta \subset \ker \omega_\Delta$ to produce the real starting point for this proof, which is

$$|\text{im } \omega_\Delta| \leq \tfrac{1}{2}|U_\Delta/H_\Delta|. \tag{4.5}$$

Let p and q be distinct odd prime numbers. We divide the analysis into two cases.

Case 1. $(p^*/q) = 1$. We are to prove that $(q/p) = 1$.

Let $\Delta = 4p^* \equiv 4 \pmod{16}$. By Definition 3.2 of H_Δ, the map T: $U_\Delta/H_\Delta \to \{\pm 1\}$ given by $T(\bar{x}) = (x/p)$ is a group isomorphism. By (4.5), $T(\text{im } \omega_\Delta) = \{+1\}$.

Let $b, c \in \mathbb{Z}$ be such that $p^* = b^2 - qc$. Then $f = [q, 2b, c]$ is a gaussian form of discriminant Δ. Since f represents q we can compute $1 = T(\omega_\Delta(f)) = T(\bar{q}) = (q/p)$, as desired.

Case 2. $(p^*/q) = -1$. We are to prove that $(q/p) = -1$.

If p or $q \equiv 1 \pmod 4$, then $(p^*/q) = (p/q)$. Thus by Case 1 we have $(q/p) = (q^*/p) = -1$.

Suppose now that $p \equiv q \equiv 3 \pmod 4$. Let $\Delta = 4pq \equiv 4 \pmod{16}$. Consider the homomorphism T: $U_\Delta/H_\Delta \to \{\pm 1\} \times \{\pm 1\}$ given by $T(\bar{x}) = ((x/p), (x/q))$. Definition 3.2 shows that T is well defined and injective. Surjectivity of T is a consequence of the Chinese Remainder Theorem 2.2 of Chapter 3. Thus T is a group isomorphism. By (4.5), $|\text{im } \omega_\Delta| \leq 2$. Since $T(\omega_\Delta([-1, 0, pq])) = ((-1/p), (-1/q)) = (-1, -1)$, we can conclude that $T(\text{im } \omega_\Delta) = \{(1,1), (-1,-1)\}$.

We now need a lemma that strengthens one aspect of Theorem 3.3i.

Lemma 4.6. Let Δ be a nonzero discriminant and let p be an odd prime divisor of Δ. Let $\mathscr{C} \in \mathscr{C}\ell(\Delta)$. Let m, n be integers that are represented by $\mathscr{C}$ and are relatively prime to p.

Then $(m/p) = (n/p)$.

Proof of Lemma 4.6. Minor modification of the first two paragraphs of the proof of Theorem 3.3i. ■

Now let m be an integer that is represented by $f = [q, 0, -p]$ and is relatively prime to Δ. Using Lemma 4.6 and the fact that f represents q and $-p$ we find that $T(\omega_\Delta(f)) = ((m/p), (m/q)) = ((q/p), (-p/q)) = ((q/p), (p^*/q)) \in \{(1,1), (-1,-1)\}$. Therefore $(p^*/q) = (q/p)$. ■

Exercises

1. Let Δ be a nonzero square. It will be proved in Sections 5 and 7 that $\ker \omega_\Delta = \operatorname{im} sq_\Delta$ and that $|\mathscr{A}m\ell(\Delta)| = |U_\Delta / H_\Delta|$. Deduce that $\operatorname{im} \omega_\Delta = U_\Delta / H_\Delta$. Thus ω_Δ is surjective.
2. Write out a detailed proof of Lemma 4.6.

5. Counting Ambiguous Classes

Theorem 5.1. Let Δ be a nonzero discriminant. Then

$$|\mathscr{A}m\ell(\Delta)| = \begin{cases} \frac{1}{2}|U_\Delta/H_\Delta| & \text{if } \Delta \text{ is a nonsquare} \\ |U_\Delta/H_\Delta| & \text{if } \Delta \text{ is a square} \end{cases}.$$

We will prove Theorem 5.1 by making separate evaluations of the two sides of its equation.

Proposition 5.2. Let Δ be a nonzero discriminant and let r be the number of odd prime divisors of Δ. Then

$$|U_\Delta/H_\Delta| = \begin{cases} 2^r & \text{if } \Delta \equiv 1 \pmod 4 \text{ or } 4 \pmod{16} \\ 2^{r+1} & \text{if } \Delta \equiv 12 \pmod{16} \text{ or } 8, 16, \text{ or } 24 \pmod{32} \\ 2^{r+2} & \text{if } \Delta \equiv 0 \pmod{32} \end{cases}.$$

Proof. Let $p_1, \ldots, p_r$ be the distinct odd prime divisors of Δ and let $A = \{\pm 1\}^r$, an abelian group of order 2^r. Define a homomorphism ϕ: $U_\Delta \to A$ by the formula $\phi(\bar{x}) = ((x/p_1), \ldots, (x/p_r))$.

Let $B = \{1\}$, $\{\pm 1\}$, or U_8, an abelian group of order 1, 2, or 4, determined so that $|A \times B|$ equals the value of $|U_\Delta/H_\Delta|$ that is asserted in Proposition 5.2. Define a homomorphism ψ: $U_\Delta \to B$ as follows. If $B = \{1\}$, then of course ψ is trivial. If $B = U_8$, then $\psi(\bar{x}) = \bar{x}$. In the remaining cases, ψ is defined by its kernel: $\psi(\bar{x}) = 1$ if and only if (a) $x \equiv 1 \pmod 4$ if $\Delta \equiv 12 \pmod{16}$ or $16 \pmod{32}$; (b) $x \equiv 1$ or $7 \pmod 8$ if $\Delta \equiv 8 \pmod{32}$; (c) $x \equiv 1$ or $3 \pmod 8$ if $\Delta \equiv 24 \pmod{32}$.

A quick glance at the Definition 3.2 of H_Δ shows that H_Δ is the kernel of the homomorphism $\phi \times \psi$: $U_\Delta/H_\Delta \to A \times B$. Since ψ and each of the r projections of ϕ are surjective, the Chinese Remainder Theorem 2.2 of Chapter 3 proves that $\phi \times \psi$ is surjective. Thus $U_\Delta/H_\Delta \simeq A \times B$, which establishes the proposition. ■

The calculation of $|\mathcal{A}m\ell(\Delta)|$ is more interesting.

Definition 5.3. A binary form f is *ambiguous* iff there exists $\gamma \in \mathbf{GL}_2(\mathbb{Z})$ such that $\gamma f = f$ and $\det(\gamma) = -1$.

A form f is *special ambiguous* iff either one of the following two conditions holds.

i. $\begin{pmatrix} 1 & 0 \\ 0 & -1 \end{pmatrix} f = f$ (equivalently, f is of the form $[a, 0, c]$).

ii. $\begin{pmatrix} 1 & 0 \\ 1 & -1 \end{pmatrix} f = f$ (equivalently, f is of the form $[a, a, c]$).

Proposition 5.4. Let Δ be a nonzero discriminant.

i. Every proper equivalence class of forms of discriminant Δ that contains an ambiguous form must contain at least one special ambiguous form.

ii. All proper equivalence classes of ambiguous forms of discriminant Δ contain exactly the same number of special ambiguous forms. That number is 2 if $\Delta < 0$ or if Δ is a square. That number is 4 if $\Delta > 0$ and Δ is a nonsquare.

iii. The number of primitive special ambiguous forms of discriminant Δ is given by the following table, where r denotes the number of distinct odd prime divisors of Δ:

Δ	$[a, 0, c]$	$[a, a, c]$	Total
$\Delta \equiv 1 \pmod 4$		2^{r+1}	2^{r+1}
$\Delta \equiv 4 \pmod{16}$	2^{r+1}		2^{r+1}
$\Delta \equiv 12 \pmod{16}$	2^{r+1}	2^{r+1}	2^{r+2}
$\Delta \equiv 8, 16$, or $24 \pmod{32}$	2^{r+2}		2^{r+2}
$\Delta \equiv 0 \pmod{32}$	2^{r+2}	2^{r+2}	2^{r+3}

We shall need a simple matrix lemma.

Lemma 5.5. i. Let $\gamma \in \mathbf{GL}_2(\mathbb{Z})$ be such that $\det \gamma = -1$ and $\gamma^2 = 1$. Then there exists $T \in \mathbf{SL}_2(\mathbb{Z})$ such that $T\gamma T^{-1} = \begin{pmatrix} 1 & 0 \\ 0 & -1 \end{pmatrix}$ or $\begin{pmatrix} 1 & 0 \\ 1 & -1 \end{pmatrix}$.

ii. There does not exist $T \in \mathbf{GL}_2(\mathbb{Z})$ such that $T\begin{pmatrix} 1 & 0 \\ 0 & -1 \end{pmatrix}T^{-1} = \begin{pmatrix} 1 & 0 \\ 1 & -1 \end{pmatrix}$.

iii. If $T \in \mathbf{SL}_2(\mathbb{Z})$ commutes with either $\begin{pmatrix} 1 & 0 \\ 0 & -1 \end{pmatrix}$ or $\begin{pmatrix} 1 & 0 \\ 1 & -1 \end{pmatrix}$, then $T = \pm\begin{pmatrix} 1 & 0 \\ 0 & 1 \end{pmatrix}$.

Proof. ii and iii are easy calculations. We prove i.

Let $\gamma \in \mathbf{GL}_2(\mathbb{Z})$ satisfy $\gamma^2 = 1$ and $\det \gamma = -1$.

The eigenvalues of γ must lie among the roots of the equation $X^2 - 1 = 0$, which γ satisfies, and their product must equal -1. Hence the eigenvalues of γ are 1 and -1. (Alternatively, observe that $\underline{x} = \underline{w} \pm \underline{w}\gamma$ satisfies the equation $\underline{x}\gamma = \pm\underline{x}$ for every $\underline{w} \in \mathbb{Z}^2$.)

Let $\underline{u} = (p \quad q) \in \mathbb{Z}^2$ satisfy $\underline{u}\gamma = \underline{u}$ and $\mathrm{GCD}(p, q) = 1$. Choose $\underline{v} = (r \quad s) \in \mathbb{Z}^2$ so that $T = \begin{pmatrix} \underline{u} \\ \underline{v} \end{pmatrix} = \begin{pmatrix} p & q \\ r & s \end{pmatrix} \in \mathbf{SL}_2(\mathbb{Z})$. Since $\underline{v} + \underline{v}\gamma$ is an eigenvector of γ for the eigenvalue 1, it must be a multiple of $\underline{u}$, say $\underline{v} + \underline{v}\gamma = a\underline{u}$, where $a \in \mathbb{Z}$. After replacing $\underline{v}$ by $\underline{v} + n\underline{u}$ with suitable $n \in \mathbb{Z}$ (which does not change the determinant of T), we can suppose that $a = 0$ or 1.

Then

$$T\gamma T^{-1} = \begin{pmatrix} \underline{u} \\ a\underline{u} - \underline{v} \end{pmatrix} T^{-1} = \begin{pmatrix} 1 & 0 \\ a & -1 \end{pmatrix} TT^{-1} = \begin{pmatrix} 1 & 0 \\ a & -1 \end{pmatrix}. \qquad \blacksquare$$

Proof of Proposition 5.4. Let $\mathscr{X} = \left\{\begin{pmatrix} 1 & 0 \\ 0 & -1 \end{pmatrix}, \begin{pmatrix} 1 & 0 \\ 1 & -1 \end{pmatrix}\right\}$.

i. Let $\mathscr{C}$ be a proper equivalence class of forms of discriminant Δ and let $f \in \mathscr{C}$ and $\gamma \in \mathbf{GL}_2(\mathbb{Z})$ satisfy $\gamma f = f$ and $\det \gamma = -1$. By Theorem 5.6 of Chapter 4 and Lemma 5.5i there exists $T \in \mathbf{SL}_2(\mathbb{Z})$ such that $T\gamma T^{-1} \in \mathscr{X}$. Then $Tf \in \mathscr{C}$ is special ambiguous, as is shown by the calculation $T\gamma T^{-1} \cdot Tf = Tf$.

ii. Let f be an ambiguous form of discriminant Δ and let $\mathscr{C}$ be its proper equivalence class. Let $G = \mathscr{A}ut^+(f) = \{\tau \in \mathbf{SL}_2(\mathbb{Z}) | \tau f = f\}$ and let $H = \{\tau^2 | \tau \in G\}$. Since G is a commutative group, H is a subgroup. We will show that the number of special ambiguous forms in $\mathscr{C}$ is equal to the index $(G : H)$ of H in G. By Corollary 5.5 of Chapter 4, if $\Delta < 0$ or if Δ is a square, then G is cyclic of order 2, 4, or 6 and so $(G : H) = 2$. If Δ is a positive nonsquare, then $G \simeq \{\pm 1\} \times \mathbb{Z}$ and so $(G : H) = 4$.

Let γ be a fixed improper automorphism of f. Let $\tau \in G$. Since $\gamma\tau$ is an improper automorphism of f, there exists $T \in \mathbf{SL}_2(\mathbb{Z})$ such that $A = T\gamma\tau T^{-1}$

$\in \mathscr{X}$. The form Tf is a special ambiguous form in $\mathscr{C}$. If $S \in \mathbf{SL}_2(\mathbb{Z})$ and $B = S\gamma\tau S^{-1} \in \mathscr{X}$, then by Lemma 5.5ii we must have $A = B$. Thus $ST^{-1} \cdot A \cdot (ST^{-1})^{-1} = S\gamma\tau S^{-1} = B = A$. By Lemma 5.5iii, $S = \pm T$ and so $Tf = Sf$. We have proved that there is a well-defined map ϕ from G to the set of special ambiguous forms in $\mathscr{C}$ given by $\phi(\tau) = Tf$.

The map ϕ is surjective. Let g be a special ambiguous form in $\mathscr{C}$. Let $A \in \mathscr{X}$ be such that $Ag = g$ and let $T \in \mathbf{SL}_2(\mathbb{Z})$ be such that $Tf = g$. Then $\tau = \gamma^{-1}T^{-1}AT$ is such that $\phi(\tau) = Tf = g$.

The map ϕ is constant on cosets of H. Let $\tau, \sigma \in G$. Let $T \in \mathbf{SL}_2(\mathbb{Z})$ be such that $A = T\gamma\tau T^{-1} \in \mathscr{X}$. Then $T\sigma \cdot \gamma\tau\sigma^2 \cdot (T\sigma)^{-1} = A$ (using Theorem 5.6 of Chapter 4). Thus $\phi(\tau) = Tf = T\sigma f = \phi(\tau\sigma^2)$.

The map ϕ is injective on cosets of H. For suppose that $\phi(\tau) = \phi(\sigma)$ where $\tau, \sigma \in G$. Let $T, S \in \mathbf{SL}_2(\mathbb{Z})$ be such that both $A = T\gamma\tau T^{-1}$ and $B = S\gamma\sigma S^{-1}$ lie in $\mathscr{X}$. Both A and B are automorphisms of $\phi(\tau)$. Since no form of nonzero discriminant can have both $\begin{pmatrix} 1 & 0 \\ 0 & -1 \end{pmatrix}$ and $\begin{pmatrix} 1 & 0 \\ 1 & -1 \end{pmatrix}$ as automorphisms, we must have $A = B$. Noting that $Tf = Sf$ implies that $S^{-1}T \in G$, we compute that $\tau = \gamma(T^{-1}S)\gamma\sigma(S^{-1}T) = \sigma(S^{-1}T)^2$. Hence τ and σ are in the same H-coset of G.

In summary, ϕ gives a bijection between the set of cosets of H in G and the set of special ambiguous forms in $\mathscr{C}$. This proves what we wanted.

iii. We first count the primitive forms of discriminant Δ that are of type $[a, 0, c]$. There are clearly none if Δ is odd, so suppose that $\Delta \equiv 0 \pmod 4$. We must have $ac = -\Delta/4$ and $\mathrm{GCD}(a, c) = 1$. The integer a is determined by specifying its sign and its set of prime divisors, which can be any subset of the set of prime divisors of $\Delta/4$. Hence the number of primitive $[a, 0, c]$ equals 2^{t+1}, where t denotes the number of prime divisors of $\Delta/4$.

We next count the primitive $[a, a, c]$ of discriminant Δ. If Δ is odd, then every divisor a of Δ leads to an integral form $[a, a, c]$, because then $c = (a^2 - \Delta)/4a \in \mathbb{Z}$. From $\Delta/a = a - 4c$ we see that $\mathrm{GCD}(a, c) = \mathrm{GCD}(a, \Delta/a)$. Thus the a that lead to primitive forms are determined by specifying their signs and their sets of prime divisors, which can be any subset of the set of r prime divisors of Δ. There are 2^{r+1} such a in all.

Now suppose that Δ is even. If $[a, a, c]$ is primitive of discriminant Δ, then a is even and c is odd. If $a = 2x$ with x odd, then $\Delta = a^2 - 4ac \equiv 12 \pmod{16}$. If $4|a$, then $\Delta = a^2 - 4ac \equiv 0 \pmod{32}$. Hence there can be no primitive forms $[a, a, c]$ unless $\Delta \equiv 12 \pmod{16}$ or $0 \pmod{32}$.

Suppose that $\Delta \equiv 12 \pmod{16}$. We are to count the primitive forms $[2x, 2x, c]$ of discriminant Δ. Primitivity implies that c is odd. Thus the equation $\Delta/4x = x - 2c$ shows that $\mathrm{GCD}(2x, c) = \mathrm{GCD}(x, \Delta/4x)$. So x is determined by its sign and its prime divisors, which lie among the r prime divisors of $\Delta/4$. There are 2^{r+1} such x in all.

Finally suppose that $\Delta \equiv 0 \pmod{32}$. We count the primitive $[4x, 4x, c]$ of discriminant Δ. From $\Delta/16x = x - c$ we learn that $\mathrm{GCD}(4x, c) = \mathrm{GCD}(x, \Delta/16x)$. We can specify x by its sign and its prime divisors, which lie among the $r + 1$ prime divisors of $\Delta/16$. There are 2^{r+2} such x in all. ■

Proof of Theorem 5.1. Combine Propositions 5.2 and 5.4. Just remember that if $\Delta < 0$, then only half the primitive forms counted in Proposition 5.4iii are gaussian, since $[a, b, c]$ is positive definite if and only if a is positive. ■

Exercises

1. Let Δ be a negative discriminant. Show that a reduced positive definite form of discriminant Δ is ambiguous if and only if it is of one of the three forms $[a, 0, c]$, $[a, a, c]$, or $[a, b, a]$.

2. Let Δ be a nonzero discriminant.

i. Show that the set of special ambiguous forms of discriminant Δ is partitioned into pairs of properly equivalent forms of the three types $\{[a, 0, c], [c, 0, a]\}$, $\{[a, a, c], [4c - a, 4c - a, c]\}$, and $\{[2c, 2c, c], [c, 0, c]\}$.

ii. Suppose that $\Delta < 0$. Prove that two forms from distinct pairs of i cannot be properly equivalent. (*Hint*: Do this first for pairs of positive definite forms by showing that two positive definite forms from distinct pairs are properly equivalent to distinct reduced forms. The reduced form for a pair of the second type may equal $[c, \pm(2c - a), c]$.) Hence give a new proof of Proposition 5.4ii for negative Δ.

3. Deduce Proposition 5.4i for positive nonsquare discriminants Δ from Exercise 6.11 of Chapter 4.

4. Let p be a prime number congruent to 1 mod 4. Let $t = [\sqrt{p}\,]$. Deduce from Theorem 5.1 that $[1, 2t, t^2 - p]$ and $[-1, 2t, p - t^2]$ are properly equivalent. Conclude that the negative Pell equation $X^2 - pY^2 = -1$ has an integral solution (by Exercise 5.4ii of Chapter 4). For an application to the equation $X^2 + Y^2 = p$, see Exercise 6.8ii of Chapter 4.

6. The Ternary Form $Y^2 - XZ$

We present one theorem from the theory of ternary quadratic forms. It will be needed for our proof of Gauss's Duplication Theorem, which will be given in the next section.

The definitions we need are straightforward analogues of the definitions from binary forms we have long been working with. We write $\mathbf{GL}_3(\mathbb{Z})$ to denote the group of 3×3 matrices with integer entries and of determinant equal to ± 1.

Definition 6.1. An *integral ternary quadratic* form is a homogeneous polynomial F of degree 2 in three variables with integer coefficients.

The *matrix* $M(F)$ of $F = aX^2 + bY^2 + cZ^2 + uXY + vYZ + wXZ$ is the symmetric 3×3 matrix

$$M = \begin{pmatrix} a & u/2 & w/2 \\ u/2 & b & v/2 \\ w/2 & v/2 & c \end{pmatrix}. \tag{6.2}$$

The *determinant* $\delta(F)$ of a ternary quadratic form F is defined to be the determinant of the matrix $M(F)$ of F.

Two integral ternary quadratic forms F and G are said to be *equivalent* iff there exists a matrix $\gamma \in \mathbf{GL}_3(\mathbb{Z})$ such that $F((X, Y, Z)\gamma) = G(X, Y, Z)$ (equivalently, such that $\gamma \cdot M(F) \cdot {}^t\gamma = M(G)$).

Theorem 6.3. Every integral ternary quadratic form of determinant $-1/4$ is equivalent to the form $Y^2 - XZ$.

We will need some simple facts about cofactor matrices, which we discuss now as they will also be useful in Sections 7 and 8. Let M be a nonsingular 3×3 matrix of determinant $\delta \neq 0$. Write $\overline{M}$ for the matrix of cofactors of M. The equation $M \cdot {}^t\overline{M} = \delta I$ shows that $\det \overline{M} = \delta^2$. Thus $\overline{M} \cdot {}^t\overline{\overline{M}} = \delta^2 I$, from which we get $\overline{\overline{M}} = \delta M$. Finally, note that the cofactor matrix is multiplicative. The cofactor matrix of $M \cdot N$ equals $\overline{M} \cdot \overline{N}$ for every pair of nonsingular 3×3 matrices M and N.

The proof of Theorem 6.3 will be based on a general lemma.

Lemma 6.4. Every integral ternary quadratic form of determinant $\delta \neq 0$ is equivalent to a form $F = aX^2 + uXY + bY^2 + wXZ + vYZ + cZ^2$ such that (a) $|a| \leq \sqrt{|u^2 - 4ab|/3}$ and (b) $|u^2 - 4ab| \leq \sqrt{64|a\delta|/3}$.

For such a form F we have the inequality $|a| \leq \frac{4}{3}\sqrt[3]{|\delta|}$.

Proof. This lemma will be seen to be a consequence of the reduction theory of binary quadratic forms.

We first interpret the second inequality (b) of the lemma, revealing a symmetry between the two inequalities (a) and (b).

We work with a 3×3 matrix

$$M = \begin{pmatrix} a & u/2 & w/2 \\ u/2 & b & v/2 \\ w/2 & v/2 & c \end{pmatrix}$$

of determinant δ as in (6.2). Let

$$\overline{M} = \begin{pmatrix} A & U/2 & W/2 \\ U/2 & B & V/2 \\ W/2 & V/2 & C \end{pmatrix}$$

be the matrix of cofactors of M. The equation $\overline{\overline{M}} = \delta M$ shows that $\delta a = BC - V^2/4$. Since $C = ab - u^2/4$, the inequality (b) is the same as (b'): $|C| \leq \sqrt{|V^2 - 4BC|/3}$.

There is one last preliminary observation to be made. Let $\gamma \in \mathbf{GL}_3(\mathbb{Z})$. Then the matrix of cofactors of $\gamma \cdot M \cdot {}^t\gamma$ equals $\overline{\gamma} \cdot \overline{M} \cdot {}^t\overline{\gamma}$, where $\overline{\gamma} = \det(\gamma) \cdot {}^t\gamma^{-1} \in \mathbf{GL}_3(\mathbb{Z})$.

Let F_0 be an integral ternary form of discriminant $\delta \neq 0$ and with matrix M_0. We use lower- and uppercase letters as before, subscripted in a natural way, to denote the entries of M_0 and $\overline{M}_0$. If the inequality (a) does not hold for M_0, choose $\theta \in \mathbf{SL}_2(\mathbb{Z})$ (by Proposition 8.3 of Chapter 2, Theorem 6.3i of Chapter 4, and Exercises 8.8 and 8.9 of Chapter 2) such that $\theta[a_0, u_0, b_0] = [a_1, u_1, b_1]$ satisfies

$$|a_1| \leq \sqrt{|u_1^2 - 4a_1b_1|/3} = \sqrt{|u_0^2 - 4a_0b_0|/3} < |a_0|.$$

Let

$$\gamma_1 = \begin{pmatrix} \theta & & \begin{matrix} 0 \\ 0 \end{matrix} \\ 0 & 0 & 1 \end{pmatrix} \in \mathbf{GL}_3(\mathbb{Z})$$

and let $M_1 = \gamma_1 \cdot M_0 \cdot {}^t\gamma_1$. Observe, in an obvious notation, that $C_1 = C_0$.

If the inequality (b') does not hold for $\overline{M}_1$, proceed similarly. Choose $\theta \in \mathbf{SL}_2(\mathbb{Z})$ such that $\theta[B_1, V_1, C_1] = [B_2, V_2, C_2]$ satisfies

$$|C_2| \leq \sqrt{|V_2^2 - 4B_2C_2|/3} < |C_1|.$$

Let

$$\gamma_2 = \begin{pmatrix} 1 & 0 \quad 0 \\ \begin{matrix} 0 \\ 0 \end{matrix} & \theta \end{pmatrix} \in \mathbf{GL}_3(\mathbb{Z})$$

and let $M_2 = \overline{\gamma}_2 \cdot M_1 \cdot {}^t\overline{\gamma}_2$, so that $\overline{M}_2 = \gamma_2 \cdot \overline{M}_1 \cdot {}^t\gamma_2$. Clearly $a_2 = a_1$.

Iterate in alternating order the two processes of the previous two paragraphs to construct a sequence $M_0, M_1, M_2, M_3, \ldots$. As long as one of the inequalities (a) or (b′) fails for M_i, the sequence can be continued to M_{i+1}. We have seen that $|a_0| > |a_1| = |a_2| > |a_3| = \cdots \geq 0$ and that $|C_0| = |C_1| > |C_2| = |C_3| > \cdots \geq 0$. Since the a_i are integers and the C_i are quarter integers, the sequence must stop with a matrix M that satisfies both (a) and (b′). The form F whose matrix is M is a form that is equivalent to F_0 and satisfies (a) and (b) as sought.

The inequality $|a| \leq \frac{4}{3}\sqrt[3]{|\delta|}$ follows directly from (a) and (b). ■

Proof of Theorem 6.3. By Lemma 6.4 every ternary form of determinant $-1/4$ is equivalent to a form with $a = u = 0$. A general form of this type has matrix

$$M = \begin{pmatrix} 0 & 0 & w/2 \\ 0 & b & v/2 \\ w/2 & v/2 & c \end{pmatrix}.$$

Let

$$\gamma = \begin{pmatrix} 1 & 0 & 0 \\ \alpha & 1 & 0 \\ \beta & 0 & 1 \end{pmatrix} \in \mathbf{GL}_3(\mathbb{Z}).$$

Let

$$M_1 = \gamma \cdot M \cdot {}^t\gamma = \begin{pmatrix} 0 & 0 & w/2 \\ \cdot & b & (v + w\alpha)/2 \\ \cdot & \cdot & c + w\beta \end{pmatrix}.$$

In our case, $\det M = -bw^2/4 = -1/4$, which implies that $b = 1$ and that $w = \pm 1$. Thus a suitable choice of $\alpha, \beta \in \mathbb{Z}$ gives

$$M_1 = \begin{pmatrix} 0 & 0 & \pm 1/2 \\ 0 & 1 & 0 \\ \pm 1/2 & 0 & 0 \end{pmatrix},$$

the matrix of the form $Y^2 \pm XZ$. The sign of the coefficient of XZ can be adjusted by transforming with $\zeta = \operatorname{diag}(1, 1, -1) \in \mathbf{GL}_3(\mathbb{Z})$. ■

Exercises

1. Let $\delta \in \mathbb{R}$. Show that there is an integral ternary quadratic form of determinant δ if and only if $4\delta \in \mathbb{Z}$.

2. Let f be an integral ternary form of determinant δ. Show that f represents an integer m such that $|m| \leq \frac{4}{3}\sqrt[3]{|\delta|}$.

3. Show that every integral ternary quadratic form with determinant 0 is equivalent to a binary form $aX^2 + uXY + bY^2$.

4. Prove that every integral ternary quadratic form of determinant 1/2 is equivalent to $XZ - 2Y^2$ or to $X^2 + XY + Y^2 + XZ + YZ + Z^2 = 1/4(X - Y)^2 + 2/3(X + Y + Z)^2 + 1/12(X + Y - 2Z)^2$ and that these two forms are not equivalent.

5. An integral ternary quadratic form is *classically integral* iff all the entries in its matrix are integers. Show that every form that is equivalent to a classically integral form is classically integral.

6. Prove that every classically integral ternary form of determinant 1 is equivalent to $2XZ - Y^2$ or to $X^2 + Y^2 + Z^2$ and that these two forms are not equivalent. To which of these forms is $X^2 - Y^2 - Z^2$ equivalent?

7. Let $m > 0$ be congruent to 1, 2, 3, 5, or 6 (mod 8). Prove that m is represented by the form $X^2 + Y^2 + Z^2$. (*Sketch of solution*: Show that there exists a classically integral form $mX^2 + bY^2 + cZ^2 + 2XZ + 2fYZ = 1/m(mX + Z)^2 + 1/b(bY + fZ)^2 + (\delta/mb)Z^2$ with $b > 0$ and determinant $\delta = mbc - mf^2 - b = 1$ that represents m. To do this, use Dirichlet's Theorem on Primes in Arithmetic Progressions. If $m \equiv 2 \pmod 4$ take b prime, $b \equiv m - 1 \pmod{4m}$. If $m \equiv 1 \pmod 4$ take $b = 2p$ with p prime, $p \equiv (3m - 1)/2 \pmod{4m}$. If $m \equiv 3 \pmod 8$ take $b = 2p$ with p prime, $p \equiv (m - 1)/2 \pmod{4m}$. Appeal to Exercise 6.)

7. The Duplication Theorem

In this section we complete the proof of the Duplication Theorem 3.8.

Proposition 7.1. Let Δ be a nonzero discriminant. Let f be a gaussian form of discriminant Δ that lies in the principal genus of $\mathcal{C}\ell(\Delta)$.

i. There exist $x, y \in \mathbb{Z}$ such that $f(x, y) \equiv 1 \pmod{\Delta}$.

ii. There exists

$$\gamma = \begin{pmatrix} r_1 & r_2 & r_3 \\ s_1 & s_2 & s_3 \\ t_1 & t_2 & t_3 \end{pmatrix} \in \mathbf{GL}_3(\mathbb{Z})$$

such that

$$f(X, Y) = (r_2X + s_2Y)^2 - (r_1X + s_1Y)(r_3X + s_3Y)$$

and

$$\mathrm{GCD}(r_1s_2 - r_2s_1, 2\Delta) = 1.$$

iii. The form f represents a square m^2 such that $m \neq 0$ is relatively prime to Δ.

Proof. i. We first prove a stronger result for the principal form f_0. Let m be an integer relatively prime to Δ such that $\overline{m} \in H_\Delta$. We will show that for every prime number p that divides Δ and every $n > 0$ there exist integers x, y such that $f_0(x, y) \equiv m \pmod{p^n}$. For an odd prime p this is possible with $y = 0$, for since $f_0(X, 0) = X^2$ and $(m/p) = 1$ we can apply Proposition 2.5i of Chapter 3. Now suppose that $p = 2$ and $\Delta \equiv 0 \pmod 4$. We want to solve the congruence $X^2 - (\Delta/4)Y^2 \equiv m \pmod{2^n}$. If $\Delta/4 \equiv 1 \pmod 4$ and $m \equiv 3 \pmod 4$, choose $X \in \{0, 2\}$ such that $X^2 - m \equiv \Delta/4 \pmod 8$ and choose Y by Proposition 2.5ii of Chapter 3. Otherwise, choose $Y \in \{0, 1, 2\}$ such that $(\Delta/4)Y^2 + m \equiv 1 \pmod 8$ and choose X by Proposition 2.5ii of Chapter 3. It follows by Proposition 2.4 of Chapter 3 that there exist $x, y \in \mathbb{Z}$ such that $f_0(x, y) \equiv m \pmod{\Delta}$.

Now let f be as in the statement of the proposition and let n be an integer relatively prime to Δ that is represented by the proper equivalence class $\mathscr{C}$ to which f belongs. Let $m \in \mathbb{Z}$ be such that $mn \equiv 1 \pmod{\Delta}$. Since f is in the principal genus we have $n \in H_\Delta$ and thus also $m \in H_\Delta$. By the preceding paragraph the principal class $\mathscr{C}_0$ represents an integer M that is congruent to m mod Δ. Hence $\mathscr{C} = \mathscr{C}_0\mathscr{C}$ represents Mn. This means that f represents Mn, which is congruent to $1 \pmod{\Delta}$.

ii. Let $f = [a, b, c]$. There exist integers l, m, n such that $\det A = -1/4$, where

$$A = \begin{pmatrix} a & b/2 & l/2 \\ b/2 & c & m/2 \\ l/2 & m/2 & n \end{pmatrix}.$$

This follows readily from Proposition 7.1i and the computation $\det A = -1/4(f(m, -l) + n\Delta)$.

Let M be the matrix of $Y^2 - XZ$:

$$M = \begin{pmatrix} 0 & 0 & -1/2 \\ 0 & 1 & 0 \\ -1/2 & 0 & 0 \end{pmatrix}.$$

By Theorem 6.3 there exists $\tau \in \mathbf{GL}_3(\mathbb{Z})$ such that $\tau \cdot M \cdot {}^t\tau = A$. Write

$$\tau = \begin{pmatrix} r_1 & r_2 & r_3 \\ s_1 & s_2 & s_3 \\ t_1 & t_2 & t_3 \end{pmatrix}.$$

Compute

$$\begin{aligned} f(X, Y) &= (X, Y, 0) \cdot A \cdot {}^t(X, Y, 0) \\ &= (X, Y, 0)\tau \cdot M \cdot {}^t((X, Y, 0)\tau) \\ &= (r_2 X + s_2 Y)^2 - (r_1 X + s_1 Y)(r_3 X + s_3 Y). \end{aligned} \tag{7.2}$$

This is nearly the result we want. But if the minor

$$\begin{vmatrix} r_1 & r_2 \\ s_1 & s_2 \end{vmatrix}$$

is not relatively prime to 2Δ, we have to modify τ. We will replace τ by $\tau' = \tau\sigma$, where $\sigma \in \mathbf{SL}_3(\mathbb{Z})$ is a suitable automorphism of $Y^2 - XZ$, i.e., σ satisfies the equation $\sigma \cdot M \cdot {}^t\sigma = M$.

For $g = \begin{pmatrix} \alpha & \beta \\ \gamma & \delta \end{pmatrix} \in \mathbf{SL}_2(\mathbb{Z})$, let

$$\sigma_g = \begin{pmatrix} \alpha^2 & \alpha\beta & \beta^2 \\ 2\alpha\gamma & \alpha\delta + \beta\gamma & 2\beta\delta \\ \gamma^2 & \gamma\delta & \delta^2 \end{pmatrix}. \tag{7.3}$$

A computation shows that $\det \sigma_g = 1$ and that σ_g is an automorphism of $Y^2 - XZ$.

Let $\tau' = \tau\sigma_g$, where g remains to be chosen, and denote the entries of τ' by r_i', s_i', t_i' to match the notation for the entries of τ. The minor

$$\begin{vmatrix} r_1' & r_2' \\ s_1' & s_2' \end{vmatrix}$$

is the final entry in the cofactor matrix

$$\bar{\tau}' = \bar{\tau}\bar{\sigma}_g = \bar{\tau} \begin{pmatrix} \cdot & \cdot & \gamma^2 \\ \cdot & \cdot & -\alpha\gamma \\ \cdot & \cdot & \alpha^2 \end{pmatrix}.$$

Thus $r_1's_2' - r_2's_1' = T_1\gamma^2 - T_2\alpha\gamma + T_3\alpha^2$, where the T_i are the bottom row entries of $\bar{\tau}$. Computing $\det \tau = \pm 1$ by expansion across the third row of τ, we see that $\mathrm{GCD}(T_1, T_2, T_3) = 1$. By Lemma 2.4 we can choose α, γ, and g

such that $\mathrm{GCD}(r_1's_2' - r_2's_1', 2\Delta) = 1$. The calculation (7.2) works for τ' as well as for τ, since $\tau' \cdot M \cdot {}^t\tau' = A$. The proof of Proposition 7.1ii is complete.

iii. $f(-s_1, r_1) = (r_1s_2 - r_2s_1)^2$, where r_1, r_2, s_1, s_2 are as in Proposition 7.1ii. ■

Proof of Theorem 3.8. Let $\mathscr{C} \in \mathscr{C}\ell(\Delta)$ be such that $\omega_\Delta(\mathscr{C}) = 1$. By Proposition 7.1iii the class $\mathscr{C}$ contains a form $[m^2, b, c]$ where m is a positive integer that is relatively prime to Δ. Since $\mathrm{GCD}(m, b) = 1$, the form $[m, b, mc]$ is gaussian. Clearly, $\mathscr{C} = \mathscr{D}^2$, where $\mathscr{D} \in \mathscr{C}\ell(\Delta)$ is the proper equivalence class of $[m, b, mc]$. Thus $\ker w_\Delta \subset \mathscr{S}q(\Delta)$. The opposite inclusion is trivial, as has already been noted. ■

Exercises

1. Let f be a gaussian form of nonzero discriminant Δ. Prove that f is in the principal genus of $\mathscr{C}\ell(\Delta)$ if and only if the congruence $f(X, Y) \equiv 1 \pmod{N}$ has a solution for every positive integer N.

2. **i.** Let $g = \begin{pmatrix} \alpha & \gamma \\ \beta & \delta \end{pmatrix} \in \mathbf{SL}_2(\mathbb{Z})$. Compute $g[X, 2Y, Z]$ and compare with the formula (7.3) for σ_g. Observe that the discriminant of $[X, 2Y, Z]$ equals $4(Y^2 - XZ)$. Hence prove without further computation that σ_g is an automorphism of $Y^2 - XZ$.

ii. Prove that $\sigma_g\sigma_h = \sigma_{gh}$ for all $g, h \in \mathbf{SL}_2(\mathbb{Z})$.

8. Sums of Three Squares

Lemma 8.1. Let $a, b, c \in \mathbb{Z}$ be such that $u = ac - b^2$ is nonzero.

i. The following two conditions are equivalent.

1. There exist integers m, n, s such that the symmetric matrix

$$R = \begin{pmatrix} a & b & m \\ b & c & n \\ m & n & s \end{pmatrix}$$

has determinant equal to 1.

2. There exist integers M, N that satisfy the system (8.2) of three congruences:

$$\begin{aligned} -a &\equiv N^2 \pmod{u}, \\ b &\equiv MN \pmod{u}, \\ -c &\equiv M^2 \pmod{u}. \end{aligned} \tag{8.2}$$

ii. If moreover $\mathrm{GCD}(a, u) = 1$ and $4 \nmid u$, then the two conditions of i are also equivalent to

3. $(a/p) = (-1/p)$ for all odd primes p that divide u.

Proof. i. $1 \Rightarrow 2$. Suppose that R exists. Let its cofactor matrix be

$$\bar{R} = \begin{pmatrix} A & B & M \\ B & C & N \\ M & N & u \end{pmatrix}.$$

The relation $\bar{\bar{R}} = R$ yields three equations that immediately imply (8.2), namely

$$\begin{aligned} a &= -N^2 + Cu, \\ b &= MN - Bu, \\ c &= -M^2 + Au. \end{aligned} \tag{8.3}$$

$2 \Rightarrow 1$. Let M, N solve the congruences (8.2). Use (8.3) to define integers A, B, and C. Next define integers m, n, and s by the equations

$$\begin{aligned} m &= BN - CM = \frac{-aM - bN}{u}, \\ n &= BM - AN = \frac{-bM - cN}{u}, \\ s &= AC - B^2 = \frac{1 - mM - nN}{u}. \end{aligned} \tag{8.4}$$

Let

$$R = \begin{pmatrix} a & b & m \\ b & c & n \\ m & n & s \end{pmatrix}.$$

It is easily checked that the last column of the cofactor matrix of R is given by

$$\bar{R} = \begin{pmatrix} \cdot & \cdot & M \\ \cdot & \cdot & N \\ \cdot & \cdot & u \end{pmatrix}.$$

Expand by minors down the third column of R and use the final equation of (8.4) to prove that $\det R = 1$.

ii. Suppose now that $\mathrm{GCD}(a, u) = 1$ and that $4 \nmid u$.

If 2 holds, then clearly $(-a/p) = 1$ for every odd prime divisor p of u. Hence $(a/p) = (-1/p)$.

Now suppose that 3 holds. By Theorem 2.6 of Chapter 3 there exists $N \in \mathbb{Z}$ such that $-a \equiv N^2 \pmod{u}$. Since $\mathrm{GCD}(N, u) = 1$, there exists $M \in \mathbb{Z}$ such that $b \equiv MN \pmod{u}$. Finally, $-N^2c \equiv ac \equiv b^2 \equiv M^2N^2 \pmod{u}$, which shows that $-c \equiv M^2 \pmod{u}$. We have found a solution to (8.2), which proves that 2 holds. ■

Lemma 8.5. Let $u \in \mathbb{Z}$ be congruent to 1 or 2 (mod 4). Let $\Delta = -4u$ and suppose that Δ is not a square.

For every $m \in U_\Delta$ there exists $x \in \ker \chi_\Delta$ such that $(x/p) = (m/p)$ for every odd prime divisor p of Δ.

In other words, the homomorphism $\phi: U_\Delta \to A$ that is defined in the proof of Proposition 5.2, is surjective on restriction to $\ker \chi_\Delta$. That is, $\phi(\ker \chi_\Delta) = A$.

Proof. Separate consideration of the three cases $\Delta \equiv 12 \pmod{16}$, $\Delta \equiv 8 \pmod{32}$, and $\Delta \equiv 24 \pmod{32}$ shows that $\ker \phi \cap \ker \chi_\Delta = H_\Delta$. Thus, with B as in the proof of Proposition 5.2, $|\phi(\ker \chi_\Delta)| = |\ker \chi_\Delta| / |\ker \phi \cap \ker \chi_\Delta| = |\ker \chi_\Delta| / |H_\Delta| = \frac{1}{2}|A \times B| = |A|$. ■

Theorem 8.6. Let u be a positive integer that is congruent to 1 or 2 (mod 4) or to 3 (mod 8). Then there exist integers x, y, z such that $x^2 + y^2 + z^2 = u$.

Proof. Suppose first that $u \equiv 1$ or $2 \pmod{4}$. Let $\Delta = -4u$. By Lemma 8.5 and Theorem 3.6 there exists $\mathscr{C} \in \mathscr{C}\ell(\Delta)$ such that $\phi(\omega_\Delta(\mathscr{C})) = \phi(-1)$. Let $[a, 2b, c] \in \mathscr{C}$ be such that $\mathrm{GCD}(a, \Delta) = 1$ and note that $ac - b^2 = u$. Since $\omega_\Delta(\mathscr{C}) = \bar{a}$, we have that $(a/p) = (-1/p)$ for every odd prime p that divides u. Thus by Lemma 8.1 there are integers m, n, s such that the matrix

$$R = \begin{pmatrix} a & b & m \\ b & c & n \\ m & n & s \end{pmatrix}$$

has determinant equal to 1. By Exercise 6.6 the ternary form

$$f = \frac{1}{a}(aX + bY + mZ)^2 + \frac{1}{au}(uY + (an - bm)Z)^2 + \frac{1}{u}Z^2$$

whose matrix is R is equivalent to the form $X^2 + Y^2 + Z^2$. Hence there exists $\gamma \in \mathbf{GL}_3(\mathbb{Z})$ such that $\gamma \cdot {}^t\gamma = R$. Taking cofactors gives $\bar{\gamma} \cdot {}^t\bar{\gamma} = \bar{R}$, which shows that the ternary form $\bar{f}$ whose matrix is $\bar{R}$ is also equivalent to $X^2 + Y^2 + Z^2$. Observe that $\bar{f}$ represents u, namely $\bar{f}(0,0,1) = u$. Since all forms in an equivalence class represent the same integers, there must exist integers x, y, and z such that $x^2 + y^2 + z^2 = u$.

Now suppose that $u \equiv 3 \pmod{8}$. The argument is similar to that of the previous case. Let $\Delta = -u$, which is an odd discriminant congruent to

5 (mod 8). Since $\chi_\Delta(-2) = 1$, there exists, by Theorem 3.6, $\mathscr{C} \in \mathscr{C}\ell(\Delta)$ such that $\omega_\Delta(\mathscr{C}) = -\bar{2}$. Let $[a', b, c'] \in \mathscr{C}$ be such that $\mathrm{GCD}(a', \Delta) = 1$. Let $a = 2a'$ and $c = 2c'$, and note that $ac - b^2 = u$. For odd primes p dividing u we calculate $(a/p) = (2/p)(a'/p) = (2/p)(-2/p) = (-1/p)$. Thus by Lemma 8.1 there are integers m, n, s such that

$$\det\begin{pmatrix} a & b & m \\ b & c & n \\ m & n & s \end{pmatrix} = 1.$$

The rest of the proof is exactly as for the case $u \equiv 1$ or 2 (mod 4). ■

The relation between Theorem 8.6 and the congruences (8.2) is similar to that between the Two Squares Theorem and the congruence $X^2 \equiv -1 \pmod 4$. Note that the proof of existence of the relevant solutions to (8.2) goes back to Theorem 3.6.

It is now a very simple matter to prove Gauss's Three Squares Theorem.

Proof of Theorem 1.1. Let $m = 4^a u$, where u is a positive integer that is not divisible by 4. We first prove that m equals a sum of three squares if and only if u equals a sum of three squares.

If $u = x^2 + y^2 + z^2$, then $m = (2^a x)^2 + (2^a y)^2 + (2^a z)^2$.

Conversely, if $m = x^2 + y^2 + z^2$ and $a \geq 1$, then x, y, and z are all even, as can be seen by consideration of the congruence $x^2 + y^2 + z^2 \equiv 0 \pmod 4$. Hence $4^{a-1}u$ and then by induction also u is a sum of three squares.

All cases except $u \equiv 7 \pmod 8$ are covered by Theorem 8.6 (or Exercise 6.7). That case is dealt with by the observation that there is no solution to the congruence $X^2 + Y^2 + Z^2 \equiv 7 \pmod 8$. ■

Gauss showed how to count the number of ways that an integer can be expressed as a sum of three squares by counting solutions of the congruences (8.2) and keeping track of their role in the proof of Theorem 8.6. We state Gauss's result without proof as the final theorem in this book.

Theorem 8.7. Let u be a positive integer, $u \neq 1, 3$, and let

$$R(u) = \left\{(x, y, z) \in \mathbb{Z}^3 \mid x^2 + y^2 + z^2 = u \text{ and } \mathrm{GCD}(x, y, z) = 1\right\}.$$

Then

$$|R(u)| = \begin{cases} 12|\mathscr{C}\ell(-4u)| & \text{if } u \equiv 1 \text{ or } 2 \pmod 4 \\ 24|\mathscr{C}\ell(-u)| & \text{if } u \equiv 3 \pmod 8 \\ 0 & \text{otherwise.} \end{cases}$$

Exercises

1. The *triangular numbers* are the integers $x(x+1)/2$, $x \geq 0 \in \mathbb{Z}$. Prove that every positive integer is the sum of three (not necessarily distinct) triangular numbers. (*Hint*: To express m as such a sum, first write $8m + 3$ as a sum of three odd squares.)

2. Deduce from Theorem 1.1 that every positive integer is the sum of four square numbers, i.e., prove that for every positive integer m there exist integers x, y, z, w such that $x^2 + y^2 + z^2 + w^2 = m$.

3. Verify Theorem 8.7 for $u = 2, 11, 17, 26, 35, 41, 51, 59$.

APPENDIX A

$\Delta = b^2 - 4ac$*

BY JEAN-PIERRE SERRE
Collège de France

The formula of the title is of course familiar; it is the *discriminant* of the quadratic polynomial $ax^2 + bx + c$.

The problem I want to discuss today is: Given an integer Δ, *what are the possible polynomials $ax^2 + bx + c$, with integer coefficients a, b, c, for which $b^2 - 4ac$ is equal to Δ?* Can we classify them?

This problem has a long history, going as far back as Gauss (circa 1800); it is not solved yet, but there have been quite exciting new results recently, as I hope to show you.

Notice first that there is an obvious necessary condition on Δ; namely Δ should be congruent to a square mod 4, i.e.,

$$\Delta \equiv 0, 1 \pmod 4).$$

Conversely, if this congruence holds, it is easy to find $a, b, c \in \mathbb{Z}$ with $\Delta = b^2 - 4ac$ (exercise). This settles the question of the *existence* of the solutions of our problem; it remains only (!) to classify them. For instance, are there some Δs for which there is a unique solution?

In this crude form, the answer is obviously "no." Indeed, the transformation $x \to x + 1$ leaves Δ invariant, but changes (a, b, c) to $(a, b + 2a, a + b + c)$. Thus, we should consider two quadratic polynomials as *equivalent* if

*Lecture organized jointly by the Singapore Mathematical Society and the Department of Mathematics, National University of Singapore, and delivered on 14 February 1985. Notes taken by Daniel E. Flath.

they differ by $x \to x + 1$, or more generally, by $x \to x + n$ ($n \in \mathbb{Z}$). But this is not enough: There are other possible transformations. To see them, it is better to use a homogeneous notation and to write our quadratic polynomials as $ax^2 + bxy + cy^2$. The transformation $x \to x + 1$ becomes $\begin{cases} x \to x + y \\ y \to y \end{cases}$, which we may write as a matrix $S = \begin{pmatrix} 1 & 1 \\ 0 & 1 \end{pmatrix}$. Since now x and y play symmetric roles, we should introduce as well the matrix $T = \begin{pmatrix} 1 & 0 \\ 1 & 1 \end{pmatrix}$, which corresponds to the transformation $\begin{cases} x \to x \\ y \to x + y \end{cases}$. And, since we can compose transformations, we should consider the group generated by S and T, which happens to be the group $\mathbf{SL}_2(\mathbb{Z})$ of two-by-two matrices $\begin{pmatrix} \alpha & \beta \\ \gamma & \delta \end{pmatrix}$, with integral coefficients and determinant 1.

Now our problem may be reformulated as follows:

Given an integer Δ, *with* $\Delta \equiv 0, 1 \pmod 4$, *classify the* $\mathbf{SL}_2(\mathbb{Z})$ *equivalence classes of quadratic forms* $ax^2 + bxy + cy^2$, *with* $a, b, c \in \mathbb{Z}$ *and* $b^2 - 4ac = \Delta$.

For the rest of this talk, we will consider *only the case where* Δ *is* < 0, i.e., equations $ax^2 + bx + c = 0$ with no real root. (The case of a positive Δ is equally interesting, but quite different, and there has been little progress on it since Gauss.) This restriction to negative Δs forces a and c to have the same sign. For convenience, we will always take them positive, and we will denote by $\underline{h}(\Delta)$ the number of such forms, modulo $\mathbf{SL}_2(\mathbb{Z})$ equivalence; we shall see below that this number is finite.

Consider a form $ax^2 + bxy + cy^2$, with $a, c > 0$, and $b^2 - 4ac = \Delta$, with $\Delta < 0$. We say that such a form is *almost reduced* if $a \leq c$ and $|b| \leq a$. *Any form can be transformed into an almost reduced one* by an element of $\mathbf{SL}_2(\mathbb{Z})$. Indeed, we can arrange that $a \leq c$ by applying the transformation $\begin{pmatrix} 0 & -1 \\ 1 & 0 \end{pmatrix}$ in case $c < a$ and we can ensure that $|b| \leq a$ by applying some shift $\begin{pmatrix} 1 & n \\ 0 & 1 \end{pmatrix}$, which leaves a invariant and replaces b by $b + 2an$. If this destroys the inequality $a \leq c$, we apply again $\begin{pmatrix} 0 & -1 \\ 1 & 0 \end{pmatrix}$, and so on. It is easily checked that this process comes to a stop after finitely many steps and gives an almost reduced form.

Theorem. The number of almost reduced forms with given discriminant $\Delta < 0$ is finite.

Proof. If $ax^2 + bxy + cy^2$ is almost reduced, we have

$$4a^2 \leq 4ac = b^2 - \Delta \leq a^2 - \Delta,$$

hence $3a^2 \leq -\Delta$; this shows that a can take only finitely many values. The same is true for b since $|b| \leq a$, and c is determined by a, b, and Δ. ∎

Corollary. $\underline{h}(\Delta)$ is finite.

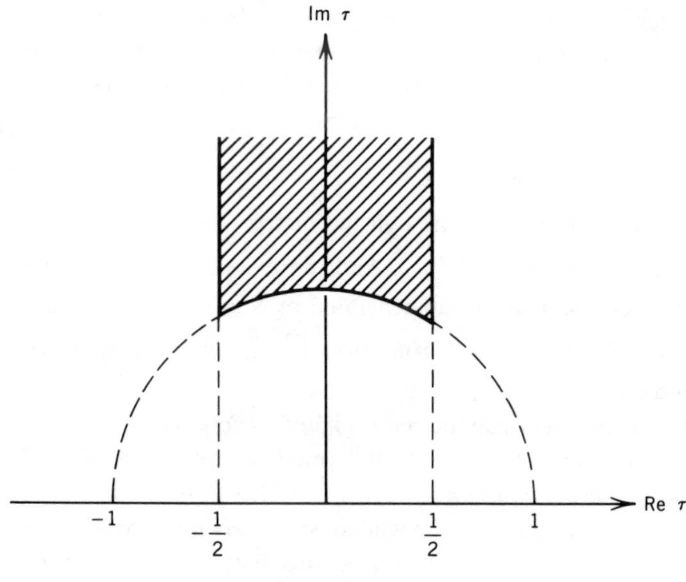

Figure A1

To go further, we need to investigate whether every $\mathbf{SL}_2(\mathbb{Z})$ equivalence class contains a *unique* almost reduced form. It turns out that this is nearly always true. I want to explain the exceptions by using a picture in the complex plane: Write $ax^2 + bxy + cy^2$ as $a(x + \tau y)(x + \bar{\tau} y)$ with some complex number τ. We may assume that $\operatorname{Im} \tau > 0$ since τ and $\bar{\tau}$ play symmetric roles. The condition $|b| \le a$ is equivalent to $|\tau + \bar{\tau}| \le 1$, that is $|\operatorname{Re} \tau| \le \frac{1}{2}$. The condition $a \le c$ translates to $\tau\bar{\tau} \ge 1$, that is $|\tau| \ge 1$. In other words, $ax^2 + bxy + cy^2$ is almost reduced precisely when τ lies in the famous shaded region pictured (boundary included) in Figure A1.

The exceptions mentioned come from the boundary. The transformation $S = \begin{pmatrix} 1 & 1 \\ 0 & 1 \end{pmatrix}$ changes τ to $\tau + 1$ relating two points on the vertical boundaries. The transformation $R = \begin{pmatrix} 0 & -1 \\ 1 & 0 \end{pmatrix}$ relates two symmetric points τ and $-1/\tau = -\bar{\tau}$ on the boundary arc.

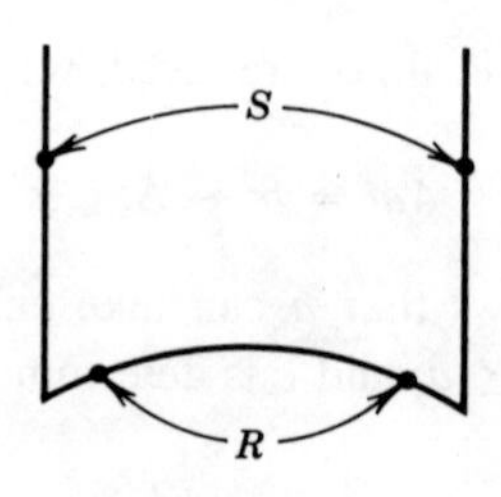

Figure A2

To get rid of the redundant almost reduced forms we throw away half the boundary. Namely:

Definition. $ax^2 + bxy + cy^2 = a(x + \tau y)(x + \bar{\tau}y)$ is *reduced* if τ lies in the region pictured in Figure A3:

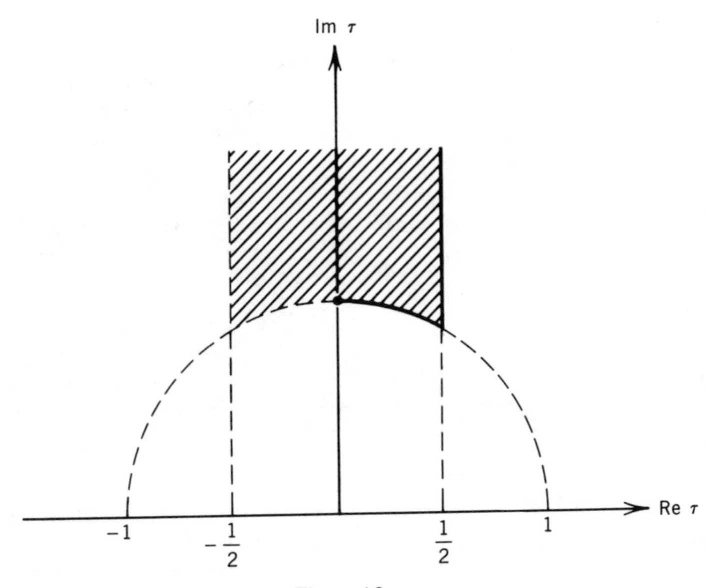

Figure A3

Equivalently, if $|b| \le a \le c$ and in case $a = |b|$, then $b = a$ and in case $a = c$, then $b \ge 0$.

This definition has been made just so that there is a *unique* reduced form in each $\mathbf{SL}_2(\mathbb{Z})$ equivalence class. Hence $\underline{h}(\Delta)$ is the number of reduced forms with discriminant Δ. This leads to a procedure for calculating $\underline{h}(\Delta)$ for a given Δ, namely listing all reduced forms as in Table A1. (The proof of the finiteness of $\underline{h}(\Delta)$ given above shows how to make this list.)

Notice that the forms $2(x^2 + xy + y^2)$ and $2(x^2 + y^2)$ of discriminants -12 and -16 are multiples of forms that appear earlier in the table under $\Delta = -3, -4$. To avoid this multiple listing we modify the game. Define a form $ax^2 + bxy + cy^2$ to be *primitive* if a, b, and c have no common factor greater than 1, and define $h(\Delta)$, the *class number* of Δ, to be the number of *primitive* reduced forms of discriminant Δ. It was a remarkable discovery of Gauss that the set C_Δ of primitive reduced forms of discriminant Δ is an abelian group in a natural way, but we shall not go into that here.*

*Call R_Δ the ring $\mathbb{Z}[\frac{1}{2}\sqrt{\Delta}\,]$ if $\Delta \equiv 0 \pmod 4$ and the ring $\mathbb{Z}[(1 + \sqrt{\Delta}\,)/2]$ if $\Delta \equiv 1 \pmod 4$. Then C_Δ is isomorphic with the "class group" $\mathrm{Pic}(R_\Delta)$ of R_Δ. When Δ is a fundamental discriminant, then R_Δ is the ring of integers of the quadratic field $\mathbb{Q}(\sqrt{\Delta}\,)$, and $h(\Delta)$ is the class number of that field.

TABLE A1

Δ	$h(\Delta)$	Reduced Forms of Discriminant Δ		
-3	1	$x^2 + xy + y^2$		
-4	1	$x^2 + y^2$		
-7	1	$x^2 + xy + 2y^2$		
-8	1	$x^2 + 2y^2$		
-11	1	$x^2 + xy + 3y^2$		
-12	2	$x^2 + 3y^2$	$2(x^2 + xy + y^2)$	
-15	2	$x^2 + xy + 4y^2$	$2x^2 + xy + 2y^2$	
-16	2	$x^2 + 4y^2$	$2(x^2 + y^2)$	
-19	1	$x^2 + xy + 5y^2$		
-20	2	$x^2 + 5y^2$	$2x^2 + 2xy + 3y^2$	
-23	3	$x^2 + xy + 6y^2$	$2x^2 - xy + 3y^2$	$2x^2 + xy + 3y^2$

TABLE A2

Δ	-3	-4	-7	-8	-11	-12	-15	-16	-19	-20
$h(\Delta)$	1	1	1	1	1	1	2	1	1	2
Δ	-23	-31	-43	-47	-59	-67	-71	-79	-83	-163
$h(\Delta)$	3	3	1	5	3	1	7	5	3	1

With computer assistance these tables have now been extended into the millions.

Looking at the tables one finds that the values $h(\Delta)$ are very irregular, but that with large $|\Delta|$, $h(\Delta)$ tends to be large as well. It has been a fundamental problem to make this last observation precise.

For technical reasons we restrict our consideration for the rest of this talk to the so-called "fundamental discriminants." A discriminant Δ is *fundamental* if it *cannot* be written $\Delta = \Delta_0 f^2$ with Δ_0 a discriminant (i.e., congruent to 0 or 1 mod 4) and f an integer greater than 1. For instance, -12 and -16 are not fundamental. This restriction is not serious because it is known how to compute all $h(\Delta)$ from the values for fundamental discriminants Δ alone.

The fundamental discriminants $\Delta < 0$ with class number $h(\Delta) = 1$ are especially interesting: They are those for which our original problem (find the quadratic equations with a given Δ) has an essentially *unique* solution. One finds easily 9 of them: $\Delta = -3, -4, -7, -8, -11, -19, -43, -67, -163$. Around 1800, Gauss conjectured that there are no more. As we shall see, this is true (but it took more than 150 years to prove).

These discriminants Δ with $h(\Delta) = 1$ have remarkable properties. Let me illustrate with the case $\Delta = -163$.

In 1772, Euler (*Mémoires de l'Académie de Berlin*, extrait d'une lettre a M. Bernoulli) discovered a curious property of the polynomial

$$x^2 + x + 41 \quad \text{(with discriminant } \Delta = -163\text{)}.$$

Namely, if you look at the table of its values for $x = 0, 1, \ldots,$

x	0	1	2	3	4	5	6	7	$\cdots$	39
$x^2 + x + 41$	41	43	47	53	61	71	83	97	$\cdots$	1601

you find only *prime numbers*, up to $x = 39$ (but $x = 40$ fails, since $40^2 + 40 + 41 = 41^2$)! The fact that this polynomial yields so many primes is *equivalent* to the equality $h(-163) = 1$. Indeed the following theorem is not hard to prove, using elementary properties of imaginary quadratic fields:

Theorem. For a prime number p that is greater than 3 and congruent to 3 mod 4, the following three properties are equivalent:

a. $h(-p) = 1$.
b. $x^2 + x + (p+1)/4$ is a prime number for every integer x such that $0 \leq x \leq (p-7)/4$.
c. $x^2 + x + (p+1)/4$ is prime for $0 \leq x < (\sqrt{p/3} - 1)/2$.

(For a proof of the equivalence (b) and (c), see, e.g., G. Frobenius, *Gesammelte Abhandlungen* III, no. 94.)

This applies to $p = 163$: By (c), it suffices to check that $x^2 + x + 41$ is prime for $x = 0, 1, 2, 3$; this *implies* it will be so up to $x = 39$.

There are other interesting facts about 163 that are related to $h(-163) = 1$. Consider for instance the transcendental number

$$e^{\pi\sqrt{163}} = 262537412640768743.99999999999925007\ldots.$$

That it is so close to being an integer can be proved a priori from $h(-163) = 1$!

[*Sketch of Proof.* One computes the value of the elliptic modular function $j(z)$ for $z = (1 + i\sqrt{163})/2$; using $h(-163) = 1$, one proves that $j(z)$ is an ordinary integer. On the other hand, the power series expansion for $j(z)$ gives:

$$\begin{aligned} j(z) &= e^{-2\pi iz} + 744 + 196884e^{2\pi iz} + \cdots \\ &= -e^{\pi\sqrt{163}} + 744 - 196884e^{-\pi\sqrt{163}} + \cdots, \end{aligned}$$

an expression in which all terms but the first two give a very small contribution (less than 10^{-12}). Hence $e^{\pi\sqrt{163}}$ is close to an integer.]

For these and other reasons, there is great interest in determining all negative fundamental discriminants Δ with class number $h(\Delta) = 1$ (or 2 or 3 or ...).

In the remainder of the talk I will review the work that has been done on this problem, some of it quite recent, some of it still in progress.

The tables suggest that the class number $h(\Delta)$ is roughly of the order of magnitude of $|\Delta|^{1/2}$. One can in fact prove readily that $h(\Delta) < 3|\Delta|^{1/2}\log|\Delta|$.

But we really want a *lower* bound for h, since we want to show that for large discriminants Δ, $h(\Delta)$ must be large as well.

Work of Gronwall in 1913 and Landau in 1918 showed that if the zeta function of $\mathbf{Q}(\sqrt{\Delta})$ has no zero between $\frac{1}{2}$ and 1, then $h(\Delta) > C|\Delta|^{1/2}/\log|\Delta|$ for a constant C which can in principle be computed. Unfortunately, the hypothesis on the zeta function has never been proved (it is a special case of GRH, the Generalized Riemann Hypothesis).

In 1934, Heilbronn completed some previous work of Deuring and proved that $\lim h(\Delta) = \infty$ when $\Delta \to -\infty$. This was soon sharpened by Siegel (1936), who showed that for every $\epsilon > 0$, there exists a positive constant C_ϵ such that $h(\Delta) \geq C_\epsilon|\Delta|^{1/2-\epsilon}$. In other words, the growth rate of $h(\Delta)$ is exactly as expected.

However, Siegel's proof gives less than might be hoped for: It is not "effective" (in plain English, the constant C_ϵ cannot be computed). The reason for this is interesting. One would like to prove that if a discriminant Δ is very large,* then $h(\Delta)$ cannot be too small. One does not know how to do that. What Siegel's proof shows, instead, is that the existence of *two* large discriminants Δ and Δ' with both $h(\Delta)$ and $h(\Delta')$ suitably small leads to a contradiction. This allows $h(\Delta)$ to be small for *one* large Δ, which is one too many!

For instance, it follows from Siegel's work that there is *at most one* fundamental discriminant Δ_{10} with class number 1 beyond the 9 previously listed as already known to Gauss. The question of the existence of Δ_{10} attained notoriety as the "problem of the tenth imaginary quadratic field."

The next progress came in 1952 when Heegner published a proof that Δ_{10} *does not exist*. However, this proof used properties of modular functions that he stated without enough justification. People could not understand his work and did not believe it (I tried myself once to follow his arguments, but got nowhere...). Hence, the question of the existence of Δ_{10} was still considered open.

*I call a negative discriminant "large" when its absolute value is large.

In 1966, Stark studied Δ_{10} in his thesis, and proved that, if it exists, it is very large: $|\Delta_{10}| > 10^{9000000}$. The following year, he succeeded in proving that Δ_{10} does not exist, thus settling the class number 1 problem. His method looked at first quite different from Heegner's; it turned out later that the two methods are closely related (and that Heegner's approach was basically correct, after all).

The same year, A. Baker also gave a solution of the class number 1 problem, by using his effective bounds for linear forms in logarithms of algebraic numbers.

With some work (by Baker himself and by Stark and Montgomery-Weinberger), this method could also be applied to $h(\Delta) = 2$, and yielded the fact that there are exactly 18 negative fundamental discriminants of class number 2, the largest being -427.

However, neither Stark's method nor Baker's applied to the problem of class number 3 or more.

To go further, we must now introduce some new objects. Recall that an *elliptic curve* E over $\mathbb{Q}$ is a nonsingular cubic

$$y^2 = x^3 + ax + b, \quad \text{with } a, b \in \mathbb{Q} \text{ and } 4a^3 + 27b^2 \neq 0.$$

To such a curve is attached a wonderful (and mysterious) analytic function $L_E(s)$, which is called its L series; it is conjectured to extend analytically to the whole $\mathbb{C}$ plane, to have a functional equation similar to the one of the Riemann zeta function (but with respect to $s \mapsto 2 - s$), etc.

This seems to have nothing to do with $h(\Delta)$. However, in 1976, Goldfeld made a startling discovery. He proved that the existence of a *single* elliptic curve E over $\mathbb{Q}$ for which $L_E(s)$ satisfies the preceding conjectures and has a zero at $s = 1$ with multiplicity at least 3 implies

$$h(\Delta) \geq C_E \log|\Delta|$$

for all* Δs, with a positive C_E that is effectively computable. (How can a hypothesis on some elliptic curve imply anything about $h(\Delta)$? Well, it is one of the many mysteries of number theory... .)

Goldfeld's theorem tells us that *if* we can find an elliptic curve E with the required properties, then $h(\Delta)$ goes to infinity effectively as $\Delta \to -\infty$. There remains the task of finding such a curve.

There are some elliptic curves, derived from modular forms and called "Weil curves," for which the holomorphy of the L series and the functional equation are known. If we choose for E such a curve, the only further

*This is correct only when $h(\Delta)$ is odd; the general statement is slightly different, see, e.g., [1].

property that is needed is that $L_E(s)$ vanish at $s = 1$ with multiplicity 3 or more. The "Birch and Swinnerton–Dyer conjecture" predicts when this should happen, namely, when the rank of the group $E(\mathbf{Q})$ of rational points of E is ≥ 3. It is easy to find such curves E. One then has to prove

$$L_E(1) = 0, \qquad L'_E(1) = 0, \qquad L''_E(1) = 0.$$

Using the functional equation of L_E (which can be fixed to have a minus sign), this reduces to proving that $L'_E(1) = 0$. But how does one show this? Of course, a computer can check that

$$L'_E(1) = 0.0000000000\ldots$$

accurate to say 10 decimal places. But that is not good enough: The theorem requires $L'_E(1)$ to be *exactly* 0.

No way around that difficulty was found for about 7 years, and as a consequence, Goldfeld's method could not be applied.

The next progress came in 1983, when Gross and Zagier found a closed formula for $L'_E(1)$. Using it, they were able to find a Weil curve E satisfying all of Goldfeld's hypotheses. The corresponding constant C_E has been computed by Oesterlé, and found to be equal to $1/7000$.

To see concretely what this means, let us apply it to the problem of determining the Δs with $h(\Delta) = 3$. Goldfeld's bound gives $|\Delta| \leq e^{21000} < 10^{9200}$. We are thus left with only a finite set of Δs to investigate. Unfortunately, that set is too large.

If the bound 10^{9200} could be brought down to 10^{2500}, one could apply a result of Montgomery–Weinberger saying that, in that range, the largest negative Δ with $h(\Delta) = 3$ is $\Delta = -907$. (Extending the Montgomery–Weinberger method is certainly possible, but would require a lot of computer work.)

Luckily, there are better elliptic curves than the one used by Gross–Zagier. Recently,* Mestre has investigated the rank 3 curve

$$y^2 + y = x^3 - 7x + 6.$$

He has been able to show that it is a Weil curve (this required computer work, too; see a recent note of his, *Comptes Rendus de l'Académie des Sciences*), and, by using the Gross–Zagier theorem, that its L series has a triple zero at $s = 1$. The corresponding C_E turns out to be $\geq 1/55$. For $h(\Delta) = 3$, this gives

$$|\Delta| \leq e^{165} < 10^{72},$$

*This work of Mestre was completed shortly after my Singapore lecture (February 1985).

which is much below Montgomery–Weinberger's 10^{2500}. The class number 3 problem is thus solved. No doubt the same method will work for other small class numbers, up to 100, say.

Of course this is not the end of the story. We would like to have effective lower bounds for $h(\Delta)$ of the size of some power of $|\Delta|$, rather than in $\log|\Delta|$. But how to get them? Will we have to wait until GRH is proved? It may take a while. . . .

References

1. Oesterlé, J., Nombres de classes des corps quadratiques imaginaires, Séminaire Nicolas Bourbaki 1983–84, *Astérisque* **121–122**, Exposé 631.
2. Zagier, D., L series of elliptic curves, the Birch–Swinnerton–Dyer conjecture, and the class number problem of Gauss, *Notices of the American Mathematical Society* **31** 739–743 (1984).

APPENDIX B
Tables

TABLE B1 The prime numbers $\leq$ 10007.

2	3	5	7	11	13	17	19	23	29
31	37	41	43	47	53	59	61	67	71
73	79	83	89	97	101	103	107	109	113
127	131	137	139	149	151	157	163	167	173
179	181	191	193	197	199	211	223	227	229
233	239	241	251	257	263	269	271	277	281
283	293	307	311	313	317	331	337	347	349
353	359	367	373	379	383	389	397	401	409
419	421	431	433	439	443	449	457	461	463
467	479	487	491	499	503	509	521	523	541
547	557	563	569	571	577	587	593	599	601
607	613	617	619	631	641	643	647	653	659
661	673	677	683	691	701	709	719	727	733
739	743	751	757	761	769	773	787	797	809
811	821	823	827	829	839	853	857	859	863
877	881	883	887	907	911	919	929	937	941
947	953	967	971	977	983	991	997	1009	1013
1019	1021	1031	1033	1039	1049	1051	1061	1063	1069
1087	1091	1093	1097	1103	1109	1117	1123	1129	1151
1153	1163	1171	1181	1187	1193	1201	1213	1217	1223
1229	1231	1237	1249	1259	1277	1279	1283	1289	1291
1297	1301	1303	1307	1319	1321	1327	1361	1367	1373
1381	1399	1409	1423	1427	1429	1433	1439	1447	1451
1453	1459	1471	1481	1483	1487	1489	1493	1499	1511
1523	1531	1543	1549	1553	1559	1567	1571	1579	1583
1597	1601	1607	1609	1613	1619	1621	1627	1637	1657
1663	1667	1669	1693	1697	1699	1709	1721	1723	1733
1741	1747	1753	1759	1777	1783	1787	1789	1801	1811
1823	1831	1847	1861	1867	1871	1873	1877	1879	1889
1901	1907	1913	1931	1933	1949	1951	1973	1979	1987
1993	1997	1999	2003	2011	2017	2027	2029	2039	2053
2063	2069	2081	2083	2087	2089	2099	2111	2113	2129
2131	2137	2141	2143	2153	2161	2179	2203	2207	2213
2221	2237	2239	2243	2251	2267	2269	2273	2281	2287
2293	2297	2309	2311	2333	2339	2341	2347	2351	2357
2371	2377	2381	2383	2389	2393	2399	2411	2417	2423
2437	2441	2447	2459	2467	2473	2477	2503	2521	2531
2539	2543	2549	2551	2557	2579	2591	2593	2609	2617
2621	2633	2647	2657	2659	2663	2671	2677	2683	2687
2689	2693	2699	2707	2711	2713	2719	2729	2731	2741
2749	2753	2767	2777	2789	2791	2797	2801	2803	2819
2833	2837	2843	2851	2857	2861	2879	2887	2897	2903
2909	2917	2927	2939	2953	2957	2963	2969	2971	2999
3001	3011	3019	3023	3037	3041	3049	3061	3067	3079
3083	3089	3109	3119	3121	3137	3163	3167	3169	3181
3187	3191	3203	3209	3217	3221	3229	3251	3253	3257
3259	3271	3299	3301	3307	3313	3319	3323	3329	3331

TABLE B1 (*Continued*)

3343	3347	3359	3361	3371	3373	3389	3391	3407	3413
3433	3449	3457	3461	3463	3467	3469	3491	3499	3511
3517	3527	3529	3533	3539	3541	3547	3557	3559	3571
3581	3583	3593	3607	3613	3617	3623	3631	3637	3643
3659	3671	3673	3677	3691	3697	3701	3709	3719	3727
3733	3739	3761	3767	3769	3779	3793	3797	3803	3821
3823	3833	3847	3851	3853	3863	3877	3881	3889	3907
3911	3917	3919	3923	3929	3931	3943	3947	3967	3989
4001	4003	4007	4013	4019	4021	4027	4049	4051	4057
4073	4079	4091	4093	4099	4111	4127	4129	4133	4139
4153	4157	4159	4177	4201	4211	4217	4219	4229	4231
4241	4243	4253	4259	4261	4271	4273	4283	4289	4297
4327	4337	4339	4349	4357	4363	4373	4391	4397	4409
4421	4423	4441	4447	4451	4457	4463	4481	4483	4493
4507	4513	4517	4519	4523	4547	4549	4561	4567	4583
4591	4597	4603	4621	4637	4639	4643	4649	4651	4657
4663	4673	4679	4691	4703	4721	4723	4729	4733	4751
4759	4783	4787	4789	4793	4799	4801	4813	4817	4831
4861	4871	4877	4889	4903	4909	4919	4931	4933	4937
4943	4951	4957	4967	4969	4973	4987	4993	4999	5003
5009	5011	5021	5023	5039	5051	5059	5077	5081	5087
5099	5101	5107	5113	5119	5147	5153	5167	5171	5179
5189	5197	5209	5227	5231	5233	5237	5261	5273	5279
5281	5297	5303	5309	5323	5333	5347	5351	5381	5387
5393	5399	5407	5413	5417	5419	5431	5437	5441	5443
5449	5471	5477	5479	5483	5501	5503	5507	5519	5521
5527	5531	5557	5563	5569	5573	5581	5591	5623	5639
5641	5647	5651	5653	5657	5659	5669	5683	5689	5693
5701	5711	5717	5737	5741	5743	5749	5779	5783	5791
5801	5807	5813	5821	5827	5839	5843	5849	5851	5857
5861	5867	5869	5879	5881	5897	5903	5923	5927	5939
5953	5981	5987	6007	6011	6029	6037	6043	6047	6053
6067	6073	6079	6089	6091	6101	6113	6121	6131	6133
6143	6151	6163	6173	6197	6199	6203	6211	6217	6221
6229	6247	6257	6263	6269	6271	6277	6287	6299	6301
6311	6317	6323	6329	6337	6343	6353	6359	6361	6367
6373	6379	6389	6397	6421	6427	6449	6451	6469	6473
6481	6491	6521	6529	6547	6551	6553	6563	6569	6571
6577	6581	6599	6607	6619	6637	6653	6659	6661	6673
6679	6689	6691	6701	6703	6709	6719	6733	6737	6761
6763	6779	6781	6791	6793	6803	6823	6827	6829	6833
6841	6857	6863	6869	6871	6883	6899	6907	6911	6917
6947	6949	6959	6961	6967	6971	6977	6983	6991	6997
7001	7013	7019	7027	7039	7043	7057	7069	7079	7103
7109	7121	7127	7129	7151	7159	7177	7187	7193	7207
7211	7213	7219	7229	7237	7243	7247	7253	7283	7297
7307	7309	7321	7331	7333	7349	7351	7369	7393	7411

TABLE B1 (*Continued*)

7417	7433	7451	7457	7459	7477	7481	7487	7489	7499
7507	7517	7523	7529	7537	7541	7547	7549	7559	7561
7573	7577	7583	7589	7591	7603	7607	7621	7639	7643
7649	7669	7673	7681	7687	7691	7699	7703	7717	7723
7727	7741	7753	7757	7759	7789	7793	7817	7823	7829
7841	7853	7867	7873	7877	7879	7883	7901	7907	7919
7927	7933	7937	7949	7951	7963	7993	8009	8011	8017
8039	8053	8059	8069	8081	8087	8089	8093	8101	8111
8117	8123	8147	8161	8167	8171	8179	8191	8209	8219
8221	8231	8233	8237	8243	8263	8269	8273	8287	8291
8293	8297	8311	8317	8329	8353	8363	8369	8377	8387
8389	8419	8423	8429	8431	8443	8447	8461	8467	8501
8513	8521	8527	8537	8539	8543	8563	8573	8581	8597
8599	8609	8623	8627	8629	8641	8647	8663	8669	8677
8681	8689	8693	8699	8707	8713	8719	8731	8737	8741
8747	8753	8761	8779	8783	8803	8807	8819	8821	8831
8837	8839	8849	8861	8863	8867	8887	8893	8923	8929
8933	8941	8951	8963	8969	8971	8999	9001	9007	9011
9013	9029	9041	9043	9049	9059	9067	9091	9103	9109
9127	9133	9137	9151	9157	9161	9173	9181	9187	9199
9203	9209	9221	9227	9239	9241	9257	9277	9281	9283
9293	9311	9319	9323	9337	9341	9343	9349	9371	9377
9391	9397	9403	9413	9419	9421	9431	9433	9437	9439
9461	9463	9467	9473	9479	9491	9497	9511	9521	9533
9539	9547	9551	9587	9601	9613	9619	9623	9629	9631
9643	9649	9661	9677	9679	9689	9697	9719	9721	9733
9739	9743	9749	9767	9769	9781	9787	9791	9803	9811
9817	9829	9833	9839	9851	9857	9859	9871	9883	9887
9901	9907	9923	9929	9931	9941	9949	9967	9973	10007

TABLE B2 Class numbers for negative discriminants. $h = |\mathscr{C}\ell(\Delta)|$, the number of proper equivalence classes of primitive positive definite integral binary quadratic forms of discriminant Δ, where $-1500 \leq \Delta \leq -3$. An asterisk (*) indicates that there is just one proper equivalence class in each genus for the given discriminant.

Δ	h	Δ	h	Δ	h	Δ	h
−3*	1	−91*	2	−179	5	−267*	2
−4*	1	−92	3	−180*	4	−268	3
−7*	1	−95	8	−183	8	−271	11
−8*	1	−96*	4	−184	4	−272	8
−11*	1	−99*	2	−187*	2	−275	4
−12*	1	−100*	2	−188	5	−276	8
−15*	2	−103	5	−191	13	−279	12
−16*	1	−104	6	−192*	4	−280*	4
−19*	1	−107	3	−195*	4	−283	3
−20*	2	−108	3	−196	4	−284	7
−23	3	−111	8	−199	9	−287	14
−24*	2	−112*	2	−200	6	−288*	4
−27*	1	−115*	2	−203	4	−291	4
−28*	1	−116	6	−204	6	−292	4
−31	3	−119	10	−207	6	−295	8
−32*	2	−120*	4	−208	4	−296	10
−35*	2	−123*	2	−211	3	−299	8
−36*	2	−124	3	−212	6	−300	6
−39	4	−127	5	−215	14	−303	10
−40*	2	−128	4	−216	6	−304	6
−43*	1	−131	5	−219	4	−307	3
−44	3	−132*	4	−220	4	−308	8
−47	5	−135	6	−223	7	−311	19
−48*	2	−136	4	−224	8	−312*	4
−51*	2	−139	3	−227	5	−315*	4
−52*	2	−140	6	−228*	4	−316	5
−55	4	−143	10	−231	12	−319	10
−56	4	−144	4	−232*	2	−320	8
−59	3	−147*	2	−235*	2	−323	4
−60*	2	−148*	2	−236	9	−324	6
−63	4	−151	7	−239	15	−327	12
−64*	2	−152	6	−240*	4	−328	4
−67*	1	−155	4	−243	3	−331	3
−68	4	−156	4	−244	6	−332	9
−71	7	−159	10	−247	6	−335	18
−72*	2	−160*	4	−248	8	−336	8
−75*	2	−163*	1	−251	7	−339	6
−76	3	−164	8	−252	4	−340*	4
−79	5	−167	11	−255	12	−343	7
−80	4	−168*	4	−256	4	−344	10
−83	3	−171	4	−259	4	−347	5
−84*	4	−172	3	−260	8	−348	6
−87	6	−175	6	−263	13	−351	12
−88*	2	−176	6	−264	8	−352*	4

TABLE B2 (*Continued*)

Δ	h	Δ	h	Δ	h	Δ	h
−355	4	−451	6	−547	3	−643	3
−356	12	−452	8	−548	8	−644	16
−359	19	−455	20	−551	26	−647	23
−360	8	−456	8	−552	8	−648	6
−363	4	−459	6	−555*	4	−651	8
−364	6	−460	6	−556	9	−652	3
−367	9	−463	7	−559	16	−655	12
−368	6	−464	12	−560	12	−656	16
−371	8	−467	7	−563	9	−659	11
−372*	4	−468	8	−564	8	−660*	8
−375	10	−471	16	−567	12	−663	16
−376	8	−472	6	−568	4	−664	10
−379	3	−475	4	−571	5	−667	4
−380	8	−476	10	−572	10	−668	11
−383	17	−479	25	−575	18	−671	30
−384	8	−480*	8	−576	8	−672*	8
−387	4	−483*	4	−579	8	−675	6
−388	4	−484	6	−580	8	−676	6
−391	14	−487	7	−583	8	−679	18
−392	8	−488	10	−584	16	−680	12
−395	8	−491	9	−587	7	−683	5
−396	6	−492	6	−588	6	−684	12
−399	16	−495	16	−591	22	−687	12
−400	4	−496	6	−592	4	−688	6
−403*	2	−499	3	−595*	4	−691	5
−404	14	−500	10	−596	14	−692	14
−407	16	−503	21	−599	25	−695	24
−408*	4	−504	8	−600	8	−696	12
−411	6	−507	4	−603	4	−699	10
−412	5	−508	5	−604	7	−700	6
−415	10	−511	14	−607	13	−703	14
−416	12	−512	8	−608	12	−704	12
−419	9	−515	6	−611	10	−707	6
−420*	8	−516	12	−612	8	−708*	4
−423	10	−519	18	−615	20	−711	20
−424	6	−520*	4	−616	8	−712	8
−427*	2	−523	5	−619	5	−715*	4
−428	9	−524	15	−620	12	−716	15
−431	21	−527	18	−623	22	−719	31
−432	6	−528	8	−624	8	−720	8
−435*	4	−531	6	−627*	4	−723	4
−436	6	−532*	4	−628	6	−724	10
−439	15	−535	14	−631	13	−727	13
−440	12	−536	14	−632	8	−728	12
−443	5	−539	8	−635	10	−731	12
−444	8	−540	6	−636	10	−732	8
−447	14	−543	12	−639	14	−735	16
−448*	4	−544	8	−640	8	−736	8

TABLE B2 (*Continued*)

Δ	*h*	Δ	*h*	Δ	*h*	Δ	*h*
−739	5	−835	6	−931	6	−1027	4
−740	16	−836	20	−932	12	−1028	16
−743	21	−839	33	−935	28	−1031	35
−744	12	−840*	8	−936	12	−1032	8
−747	6	−843	6	−939	8	−1035	8
−748	6	−844	9	−940	6	−1036	12
−751	15	−847	10	−943	16	−1039	23
−752	10	−848	12	−944	18	−1040	16
−755	12	−851	10	−947	5	−1043	8
−756	12	−852	8	−948	12	−1044	12
−759	24	−855	16	−951	26	−1047	16
−760*	4	−856	6	−952	8	−1048	6
−763	4	−859	7	−955	4	−1051	5
−764	13	−860	14	−956	15	−1052	13
−767	22	−863	21	−959	36	−1055	36
−768	8	−864	12	−960*	8	−1056	16
−771	6	−867	6	−963	6	−1059	6
−772	4	−868	8	−964	12	−1060	8
−775	12	−871	22	−967	11	−1063	19
−776	20	−872	10	−968	10	−1064	20
−779	10	−875	10	−971	15	−1067	12
−780	12	−876	12	−972	9	−1068	6
−783	18	−879	22	−975	16	−1071	20
−784	8	−880	8	−976	12	−1072	6
−787	5	−883	3	−979	8	−1075	6
−788	10	−884	16	−980	12	−1076	22
−791	32	−887	29	−983	27	−1079	34
−792	8	−888	12	−984	12	−1080	12
−795*	4	−891	6	−987	8	−1083	6
−796	9	−892	7	−988	6	−1084	11
−799	16	−895	16	−991	17	−1087	9
−800	12	−896	16	−992	16	−1088	16
−803	10	−899	14	−995	8	−1091	17
−804	12	−900	8	−996	12	−1092*	8
−807	14	−903	16	−999	24	−1095	28
−808	6	−904	8	−1000	10	−1096	12
−811	7	−907	3	−1003	4	−1099	6
−812	12	−908	15	−1004	21	−1100	12
−815	30	−911	31	−1007	30	−1103	23
−816	12	−912	8	−1008	8	−1104	16
−819	8	−915	8	−1011	12	−1107	6
−820	8	−916	10	−1012*	4	−1108	6
−823	9	−919	19	−1015	16	−1111	22
−824	20	−920	20	−1016	16	−1112	14
−827	7	−923	10	−1019	13	−1115	10
−828	6	−924	12	−1020	12	−1116	12
−831	28	−927	20	−1023	16	−1119	32
−832	8	−928*	4	−1024	8	−1120*	8

TABLE B2 (*Continued*)

Δ	h	Δ	h	Δ	h	Δ	h
−1123	5	−1219	6	−1315	6	−1411	4
−1124	20	−1220	16	−1316	24	−1412	16
−1127	24	−1223	35	−1319	45	−1415	34
−1128	8	−1224	16	−1320*	8	−1416	16
−1131	8	−1227	4	−1323	6	−1419	12
−1132	9	−1228	9	−1324	9	−1420	12
−1135	18	−1231	27	−1327	15	−1423	9
−1136	14	−1232	16	−1328	18	−1424	24
−1139	16	−1235	12	−1331	11	−1427	15
−1140	16	−1236	12	−1332	8	−1428*	8
−1143	20	−1239	32	−1335	28	−1431	30
−1144	12	−1240	8	−1336	12	−1432	6
−1147	6	−1243	4	−1339	8	−1435*	4
−1148	14	−1244	19	−1340	18	−1436	19
−1151	41	−1247	26	−1343	34	−1439	39
−1152	8	−1248*	8	−1344	16	−1440	16
−1155*	8	−1251	12	−1347	6	−1443	8
−1156	8	−1252	8	−1348	8	−1444	10
−1159	16	−1255	12	−1351	24	−1447	23
−1160	20	−1256	26	−1352	14	−1448	18
−1163	7	−1259	15	−1355	12	−1451	13
−1164	12	−1260	12	−1356	18	−1452	12
−1167	22	−1263	20	−1359	28	−1455	28
−1168	8	−1264	10	−1360	8	−1456	12
−1171	7	−1267	6	−1363	6	−1459	11
−1172	18	−1268	10	−1364	28	−1460	20
−1175	30	−1271	40	−1367	25	−1463	32
−1176	12	−1272	12	−1368	12	−1464	12
−1179	10	−1275	8	−1371	12	−1467	4
−1180	8	−1276	10	−1372	7	−1468	9
−1183	14	−1279	23	−1375	20	−1471	23
−1184	20	−1280	16	−1376	20	−1472	12
−1187	9	−1283	11	−1379	16	−1475	12
−1188	12	−1284	20	−1380*	8	−1476	16
−1191	24	−1287	20	−1383	18	−1479	28
−1192	6	−1288	8	−1384	10	−1480	12
−1195	8	−1291	9	−1387	4	−1483	7
−1196	24	−1292	12	−1388	15	−1484	24
−1199	38	−1295	36	−1391	44	−1487	37
−1200	12	−1296	12	−1392	12	−1488	8
−1203	6	−1299	8	−1395	8	−1491	12
−1204	8	−1300	12	−1396	14	−1492	10
−1207	18	−1303	11	−1399	27	−1495	20
−1208	12	−1304	22	−1400	16	−1496	28
−1211	14	−1307	11	−1403	14	−1499	13
−1212	10	−1308	12	−1404	12	−1500	10
−1215	18	−1311	28	−1407	24		
−1216	12	−1312	8	−1408	8		

TABLE B3 Negative discriminants with one class per genus. The known negative discriminants for which every genus of primitive integral binary quadratic forms contains just a single proper equivalence class.

−3	−4	−7	−8	−11	−12	−15	−16	−19	−20
−24	−27	−28	−32	−35	−36	−40	−43	−48	−51
−52	−60	−64	−67	−72	−75	−84	−88	−91	−96
−99	−100	−112	−115	−120	−123	−132	−147	−148	−160
−163	−168	−180	−187	−192	−195	−228	−232	−235	−240
−267	−280	−288	−312	−315	−340	−352	−372	−403	−408
−420	−427	−435	−448	−480	−483	−520	−532	−555	−595
−627	−660	−672	−708	−715	−760	−795	−840	−928	−960
−1012	−1092	−1120	−1155	−1248	−1320	−1380	−1428	−1435	−1540
−1632	−1848	−1995	−2080	−3003	−3040	−3315	−3360	−5280	−5460
−7392									

TABLE B4 Class numbers for positive discriminants. $h = |\mathscr{C}\ell(\Delta)|$, the number of proper equivalence classes of primitive integral binary quadratic forms of nonsquare discriminant Δ, where $5 \le \Delta < 1600$. An asterisk (*) indicates that the norm of the fundamental unit of the Δ-order $\mathcal{O}_\Delta$ equals -1.

Δ	h	Δ	h	Δ	h	Δ	h
5*	1	105	4	197*	1	285	4
8*	1	108	2	200*	2	288	4
12	2	109*	1	201	2	292*	1
13*	1	112	2	204	4	293*	1
17*	1	113*	1	205	4	296*	2
20*	1	116*	1	208	2	297	2
21	2	117	2	209	2	300	4
24	2	120	4	212*	1	301	2
28	2	124	2	213	2	304	2
29*	1	125*	1	216	2	305	4
32	2	128	2	217	2	308	2
33	2	129	2	220	4	309	2
37*	1	132	2	221	4	312	4
40*	2	133	2	224	4	313*	1
41*	1	136	4	228	2	316	6
44	2	137*	1	229*	3	317*	1
45	2	140	4	232*	2	320	4
48	2	141	2	233*	1	321	6
52*	1	145*	4	236	2	325*	2
53*	1	148*	3	237	2	328*	4
56	2	149*	1	240	4	329	2
57	2	152	2	241*	1	332	2
60	4	153	2	244*	1	333	2
61*	1	156	4	245	2	336	4
65*	2	157*	1	248	2	337*	1
68*	1	160	4	249	2	340*	2
69	2	161	2	252	4	341	2
72	2	164*	1	253	2	344	2
73*	1	165	4	257*	3	345	4
76	2	168	4	260*	2	348	4
77	2	172	2	261	2	349*	1
80	2	173*	1	264	4	352	4
84	2	176	2	265*	2	353*	1
85*	2	177	2	268	2	356*	1
88	2	180	2	269*	1	357	4
89*	1	181*	1	272	2	360	4
92	2	184	2	273	4	364	4
93	2	185*	2	276	2	365*	2
96	4	188	2	277*	1	368	2
97*	1	189	2	280	4	369	2
101*	1	192	4	281*	1	372	2
104*	2	193*	1	284	2	373*	1

TABLE B4 (*Continued*)

Δ	h	Δ	h	Δ	h	Δ	h
376	2	473	6	572	4	669	2
377	4	476	4	573	2	672	8
380	4	477	2	577*	7	673*	1
381	2	480	8	580*	4	677*	1
384	4	481*	2	581	2	680*	4
385	4	485*	2	584	4	681	2
388*	1	488*	2	585	4	684	4
389*	1	489	2	588	4	685*	2
392	2	492	4	589	2	688	2
393	2	493*	2	592	6	689	8
396	8	496	2	593*	1	692*	1
397*	1	497	2	596*	1	693	4
401*	5	500*	1	597	2	696	4
404*	3	501	2	600	4	697*	6
405	2	504	4	601*	1	700	4
408	4	505	8	604	2	701*	1
409*	1	508	2	605	2	704	4
412	2	509*	1	608	4	705	4
413	2	512	2	609	4	708	2
416	4	513	2	612	2	709*	1
417	2	516	2	613*	1	712	4
420	4	517	2	616	4	713	2
421*	1	520*	4	617*	1	716	2
424*	2	521*	1	620	4	717	2
425*	2	524	2	621	6	720	4
428	2	525	4	624	8	721	2
429	4	528	4	628*	1	724*	1
432	2	532	2	629*	2	725	4
433*	1	533*	2	632	2	728	4
436*	1	536	2	633	2	732	4
437	2	537	2	636	4	733*	3
440	4	540	4	637	2	736	4
444	4	541*	1	640	4	737	2
445*	4	544	8	641*	1	740*	2
448	4	545	4	644	2	741	4
449*	1	548*	1	645	4	744	4
452*	1	549	2	648	2	745	4
453	2	552	4	649	2	748	4
456	4	553	2	652	2	749	2
457*	1	556	2	653*	1	752	2
460	4	557*	1	656	2	753	2
461*	1	560	4	657	2	756	6
464	2	561	4	660	4	757*	1
465	4	564	6	661*	1	760	4
468	2	565*	2	664	2	761*	3
469	6	568	6	665	4	764	2
472	2	569*	1	668	2	765	4

TABLE B4 (*Continued*)

Δ	h	Δ	h	Δ	h	Δ	h
768	4	865*	2	964*	1	1060*	2
769*	1	868	2	965*	2	1061*	1
772*	1	869	2	968	2	1064	4
773*	1	872*	2	969	4	1065	4
776	4	873	2	972	2	1068	4
777	8	876	8	973	2	1069*	1
780	8	877*	1	976	2	1072	2
781	2	880	8	977*	1	1073*	2
785*	6	881*	1	980	2	1076*	3
788*	3	884	4	981	2	1077	2
789	2	885	4	984	4	1080	4
792	4	888	4	985*	6	1081	2
793	8	889	2	988	4	1084	2
796	2	892	6	989	2	1085	4
797*	1	893	2	992	4	1088	4
800	4	896	8	993	6	1092	4
801	2	897	8	996	2	1093*	5
804	2	901*	4	997*	1	1096*	4
805	4	904*	8	1000*	2	1097*	1
808*	2	905	8	1001	4	1100	4
809*	1	908	2	1004	2	1101	6
812	4	909	2	1005	4	1104	4
813	2	912	4	1008	4	1105*	4
816	4	913	2	1009*	7	1108*	1
817	10	916*	3	1012	2	1109*	1
820	4	917	2	1013*	1	1112	2
821*	1	920	4	1016	6	1113	4
824	2	921	2	1017	2	1116	4
825	4	924	8	1020	8	1117*	1
828	4	925*	2	1021*	1	1120	8
829*	1	928	4	1025*	4	1121	2
832	4	929*	1	1028*	3	1124*	1
833	2	932*	1	1029	2	1125	2
836	2	933	2	1032	4	1128	4
837	6	936	4	1033*	1	1129*	9
840	8	937*	1	1036	4	1132	2
844	2	940	12	1037*	2	1133	2
845*	2	941*	1	1040	4	1136	2
848	2	944	2	1041	2	1137	2
849	2	945	4	1044	2	1140	4
852	2	948	2	1045	8	1141	2
853*	1	949*	2	1048	2	1144	4
856	2	952	4	1049*	1	1145*	4
857*	1	953*	1	1052	2	1148	4
860	4	956	2	1053	2	1149	2
861	4	957	4	1056	8	1152	8
864	4	960	8	1057	2	1153*	1

TABLE B4 (*Continued*)

Δ	h	Δ	h	Δ	h	Δ	h
1157*	2	1253	2	1349	2	1448*	2
1160*	4	1256*	2	1352*	2	1449	4
1161	2	1257	6	1353	4	1452	4
1164	8	1260	8	1356	4	1453*	1
1165*	2	1261*	2	1357	2	1456	4
1168	2	1264	6	1360	4	1457	2
1169	2	1265	4	1361*	1	1460*	2
1172*	1	1268*	1	1364	2	1461	2
1173	4	1269	2	1365	8	1464	4
1176	4	1272	4	1368	8	1465*	2
1177	2	1273	2	1372	2	1468	2
1180	4	1276	4	1373*	3	1469	4
1181*	1	1277*	1	1376	4	1472	4
1184	4	1280	4	1377	2	1473	2
1185	4	1281	4	1380	4	1476	2
1188	2	1284	6	1381*	1	1477	2
1189*	2	1285*	2	1384*	6	1480*	4
1192*	2	1288	8	1385*	2	1481*	1
1193*	1	1289*	1	1388	2	1484	4
1196	4	1292	8	1389	2	1485	4
1197	4	1293	2	1392	4	1488	4
1200	4	1297*	11	1393	10	1489*	3
1201*	1	1300*	6	1396*	3	1492*	3
1204	2	1301*	1	1397	2	1493*	1
1205	4	1304	6	1400	4	1496	4
1208	2	1305	8	1401	2	1497	2
1209	4	1308	4	1404	4	1500	4
1212	4	1309	4	1405	4	1501	2
1213*	1	1312	8	1408	4	1504	4
1216	4	1313*	4	1409*	1	1505	4
1217*	1	1316	2	1412*	1	1508	4
1220	4	1317	2	1413	2	1509	6
1221	8	1320	8	1416	4	1512	4
1224	8	1321*	1	1417*	2	1513	4
1228	2	1324	2	1420	4	1516	2
1229*	3	1325*	2	1421	2	1517	4
1232	4	1328	2	1424	2	1520	8
1233	2	1329	2	1425	12	1524	6
1236	2	1332	6	1428	4	1525*	4
1237*	1	1333	2	1429*	5	1528	2
1240	4	1336	2	1432	2	1529	2
1241*	2	1337	2	1433*	1	1532	2
1244	2	1340	4	1436	6	1533	4
1245	4	1341	2	1437	2	1536	4
1248	8	1344	8	1440	8	1537	4
1249*	1	1345	12	1441	2	1540	4
1252*	1	1348*	1	1445*	2	1541	2

TABLE B4 (*Continued*)

Δ	*h*	Δ	*h*	Δ	*h*	Δ	*h*
1544	4	1557	2	1572	2	1585*	2
1545	4	1560	8	1573	6	1588*	1
1548	4	1561	2	1576*	2	1589	2
1549*	1	1564	4	1577	2	1592	2
1552	2	1565*	2	1580	4	1593	6
1553*	1	1568	4	1581	4	1596	16
1556*	3	1569	2	1584	8	1597*	1

Bibliography

Number theory has been blessed with many excellent books. This bibliography is a list of those that were available to me during the writing of this book. I have drawn most heavily from [5], [12], [13], [14], [31], and [33].

The reader who would learn more about rational and integral quadratic forms must begin by studying the fields of p-adic numbers. There are fine introductions in [4] and [27]. His ultimate goal must be a close study of [5], which is magnificent.

A proof of Dirichlet's Theorem on Primes in Arithmetic Progressions can be found in [27]. For a superb introduction to the distribution of prime numbers, see [28].

[1] Auslander, L. and Tolimieri, R., Ring structure and the Fourier transform, *The Mathematical Intelligencer* **7**, 49–52 (1985).

[2] Bachmann, P., *Niedere Zahlentheorie*, Chelsea, New York, 1968.

[3] Baker, A., *A Concise Introduction to the Theory of Numbers*, Cambridge University Press, Cambridge, 1984.

[4] Borevich, Z. I. and Shafarevich, I. R., *Number Theory*, Academic, New York, 1966.

[5] Cassels, J. W. S., *Rational Quadratic Forms*, Academic, London, 1978.

[6] Cohn, H., *A Second Course in Number Theory*, Wiley, New York, 1962.

[7] Davenport, H., *The Higher Arithmetic*, 5th ed., Cambridge University Press, Cambridge, 1982.

[8] Dickson, L. E., *Introduction to the Theory of Numbers*, The University of Chicago Press, Chicago, 1929.

[9] Dirichlet, P. G. L., *Vorlesungen über Zahlentheorie, herausgegeben von R. Dedekind*, 4th ed., Chelsea, New York, 1968.

[10] Edwards, H. M., *Fermat's Last Theorem*, Springer, New York, 1977.

[11] Frei, G., Leonhard Euler's Convenient Numbers, *The Mathematical Intelligencer*, **7**, 55–58, 64 (1985).

[12] Gauss, C. F., *Disquisitiones Arithmeticae*, Springer, New York, 1986.

[13] Hardy, G. H. and Wright, E. M., *An Introduction to the Theory of Numbers*, 5th ed., Oxford University Press, Oxford, 1979.

[14] Hua, L. K., *Introduction to Number Theory*, Springer, New York, 1982.

[15] Ireland, K. and Rosen, M., *A Classical Introduction to Modern Number Theory*, Springer, New York, 1982.

[16] Jones, B. W., *The Arithmetic Theory of Quadratic Forms*, The Mathematical Association of America, Providence, R.I., 1950.

[17] Knuth, D. E., *The Art of Computer Programming*, Vol. 2, 2nd ed., Addison-Wesley, Reading, Mass., 1981.

[18] Landau, E., *Elementary Number Theory*, Chelsea, New York, 1958.

[19] Landau, E., *Handbuch der Lehre von der Verteilung der Primzahlen*, Chelsea, New York, 1953.

[20] LeVeque, W. J., *Fundamentals of Number Theory*, Addison-Wesley, Reading, Mass., 1977.

[21] LeVeque, W. J., *Topics in Number Theory*, Vol. II, Addison-Wesley, Reading, Mass., 1956.

[22] Mathews, G. B., *Theory of Numbers*, *Part I*, Deighton, Bell and Co., Cambridge, 1892.

[23] Mordell, L. J., *Diophantine Equations*, Academic, London, 1969.

[24] Nagell, T., *Introduction to Number Theory*, 2nd ed., Chelsea, New York, 1981.

[25] Narkiewicz, W., *Number Theory*, World Scientific, Singapore, 1983.

[26] Niven, I. and Zuckerman, H. S., *An Introduction to the Theory of Numbers*, 4th ed., Wiley, New York, 1980.

[27] Serre, J.-P., *A Course in Arithmetic*, Springer, New York, 1973.

[28] Shapiro, H. N., *Introduction to the Theory of Numbers*, Wiley, New York, 1983.

[29] Sierpinski, W., *Elementary Theory of Numbers*, Panstwowe Wydawnictwo Naukowe, Warsaw, 1964.

[30] Stark, H. M., *An Introduction to Number Theory*, MIT Press, Cambridge, Mass., 1978.

[31] Venkov, B. A., *Elementary Number Theory*, Wolters-Noordhoff Publishing, Groningen, 1970.

[32] Vinogradov, I. M., *Elements of Number Theory*, Dover, London, 1954.

[33] Weil, A., *Number Theory: An Approach through History*, Birkhauser, Boston, 1983.

[34] Weil, A., *Number Theory for Beginners*, Springer, New York, 1979.

Subject Index

Notation Index

$\mathbb{Z}$	The ring of integers.
$\mathbb{Q}$	The field of rational numbers.
$\mathbb{R}$	The field of real numbers.
$\mathbb{C}$	The field of complex numbers.
$A^{\times}$	The group of units in a ring A.
$A[X]$	The ring of polynomials in one variable with coefficients in a ring A.
$\mathbf{GL}_n(\mathbb{R})$	The group of invertible $n \times n$ matrices with entries in $\mathbb{R}$.
$\mathbf{GL}_n(\mathbb{Z})$	The group of $n \times n$ matrices with entries in $\mathbb{Z}$ and determinant ± 1.
$\mathbf{SL}_n(\mathbb{Z})$	The group of $n \times n$ matrices with entries in $\mathbb{Z}$ and determinant $+1$.
$\operatorname{diag}(\cdot)$	The block diagonal matrix with specified blocks.
${}^t g$	The transpose of a matrix.
det, tr	The determinant and trace of a square matrix.
Im, Re	The imaginary and real parts of a complex number.
ker	The kernel of a homomorphism.
im	The image of a map.
deg	The degree of a nonzero polynomial.
iff	If and only if, used in a definition.
sign	The sign (± 1) of a nonzero real number.
$\lvert S \rvert$	The number of elements in a finite set S.
$[x]$	The greatest integer less than or equal to a real number x.